# Contents

# Acknowledgements

When two people generate a book it is a complicated process requiring much discussion and patience. We would like to thank all those who have been caught up in it. In particular those who read parts of it, as well as Carol, Alice and Meara, all of whom helped produce the various drafts. Howard Green and the staff at Kogan Page have been long-suffering in the negotiations over the final format – which we hope you will enjoy as well as find helpful.

We would also like to thank all those governors, heads and others with whom we have endlessly debated these issues over the past six years.

Finally, we wish to acknowledge the influence of some illuminating reading. If you get something from what we've written, you'll find these books worthwhile, too.

Belbin, Meredith (1992) *Management Teams, Why They Succeed or Fail*, Heinemann.

Belbin, Meredith (1994) *Team Roles at Work*, Butterworth/Heinemann.

Burgess, Tyrrell (ed.) (1992) *Accountability in Schools*, Longman.

Handy, Charles (1976) *Understanding Organisations*, Penguin.

Lawton, Denis (1992) *Education and Politics in the 1990s: Conflict or Consensus?* Falmer.

Ranson, Stewart and Tomlinson, John (eds) *School Cooperation: New Forms of Local Governance*, Longman.

Sallis, Joan (1988) *A New Approach to Accountability*, Routledge.

Thody, Angela (1992) *Moving to Management*, Fulton.

# Series Editor's Foreword

The government's educational reforms have created an unprecedented rate of change in schools. They have also raised fundamental questions about the purpose of education and the nature of school management and leadership. Similar changes are occurring worldwide.

In this context, there is an urgent need for all of us with an interest in education to step back and reflect on recent educational reforms, to reaffirm old truths and successful practice where appropriate, to sift out and implement the best of new ideas, modifying or abandoning those which are a distraction from the central purpose of schools, to ensure that an education of high quality is a guaranteed opportunity for *all* our children and young people.

This series aims to satisfy the demand for short, readable books designed for busy people and with a focus on single issues at the forefront of school management and leadership. Written by reflective practitioners who are either working in schools or directly with those who do, the series celebrates the ideals, skills and experience of professionals in education who want to see further improvements in our schools.

Government legislation over the last decade has substantially redefined the roles of the school governor and the responsibilities of governing bodies. The fact that this book has the word 'new' in its title indicates the unresolved issues still arising from these roles and responsibilities. The book is an important and ambitious attempt to provide some answers and to chart a way ahead. The authors, Ann Holt and Tom Hinds, are particularly well qualified in this field. Ann Holt has headed a unit in the Department for Education with responsibility for the recruitment of school governors and is currently the director of Christians in Education. Tom Hinds was the chief officer of a local education authority and now works as an educational consultant. Both have extensive experience of governor training and they remain actively involved in schools as governors.

'We believe in governing bodies' is a conviction stated early in the book which pervades every page. The principles of school governance

5

are clearly identified and then interwoven with helpful examples of their practical application. Heads are in a unique position to foster or to block the partnership between governors and school and they need good governors as much as the governors need an effective head to fulfil their responsibilities. It is argued that governors should have a central, not a tangential, role in securing quality in education, particularly through their involvement with policy formulation and strategic planning.

Far from keeping politics out of education, the authors maintain that there should be an essential and constructive link between politics and education, with governors playing their part in making the link a positive one. The declining influence of local education authorities has resulted in a growing vacuum between central government and schools. Government should recognize the active contribution that governors can make to the process of school improvement and should not leave them with the mundane task of ensuring that centrally determined policies are implemented at the grassroots. The strength of a participative democracy and the success of learning organizations suggest that all the stakeholders must be as fully involved as possible in helping to raise the standards of education.

There is helpful discussion about complex and contentious issues like the application of business models of management to education, the ownership of schools (whose school is it anyway?) and accountability. Having mapped the territory and analysed the issues, the authors then go on to bring the picture of school governance into focus. They stress the importance of educational values as the compass for securing a route through the rapidly changing territory. Boundaries between the stakeholders in education must not become barriers. There is an important distinction between 'being related' and 'building a relationship', particularly where these phrases refer to governors and their schools. Building this relationship must take into account practical steps like understanding how groups behave, resolving conflict and establishing ground rules for discussion.

The statement of conviction at the beginning of the book is complemented by two challenging questions at the end: can the government let governing bodies breathe independently, and can headteachers help governors to work on the boundary between their schools and the wider community? This is an impressive and clearly presented book on a topic vital for the well-being of our schools.

*Howard Green*
*Eggbuckland, August 1994*

# The New School Governor

0749410191

**To Joan Sallis**
**An original model of a thinking governor**

First published in 1994

Kogan Page Limited
120 Pentonville Road
London N1 9JN

© Ann Holt and Tom Hinds, 1994

'The Voice of Authority', copyright 1956 Kingsley Amis, is reprinted by permission of Jonathan Clowes Ltd., London, on behalf of Kingsley Amis. 'The Clarke's Tale' by Taconia Bridge first appeared in *Education Journal*, October, 1992, published by Longman Group.
*Accountability in Schools*, Tyrrell Burgess (ed.) 1992, published by Longman Group.

**British Library Cataloguing in Publication Data**

A CIP record for this book is available from the British Library.

ISBN 0 7494 1019 1

Typeset by Saxon Graphics Ltd, Derby
Printed and bound in Great Britain by Biddles Ltd, Guildford and King's Lynn.

# Introduction: Access, Autonomy and Accountability

New-style governing bodies sprang from a desire amongst parents and others to gain greater access to what was going on inside schools. Various Secretaries of State have responded to that request for access with a series of initiatives that claim to give not only access, but greater autonomy. However, like its namesake credit card, access may take the waiting out of wanting, but it doesn't pay the bill.

The new governing bodies are a sharp focus for greater public access to, and awareness of, schools. They are also the agents of greater school autonomy within schemes of local management, or self-management, and are in a new position of accountability as power is redistributed and new balances of power emerge. All of this could be to the immense good of the school if used wisely, if, like the Access card, governing bodies prove to be the flexible friends that our schools need to move more quickly and effectively on the education journey. But equally, if governing bodipes are unwise, particularly under the material pressure to take the waiting out of wanting, then serious damage may have been caused before the Local Education Authority (LEA) or a specially-convened Education Association can get to calling the account in and restoring the balance.

## Where we're coming from

The potential for good or ill of the new governing bodies is so much beyond mere access to schools that we decided that we wanted to write this book. It attempts to examine some of the whys and wherefores of governing, and to bring greater clarity than can be achieved by the temptation (into which we have all fallen) of rushing straight to the how.

We have been providing information and training to governors for more than a decade, and we are governors ourselves. We have experienced the fears and frustrations that cause initial enthusiasms to ebb away. Such fear or frustration is often born out of a lack of understanding about what it is that we have got ourselves into. Governors may be like the parent who, in 1989, stood up at a conference to take Kenneth Baker, the then Secretary of State for Education, to task: 'We didn't ask for all this', she remonstrated, 'we only wanted influence.' 'Ah', replied Mr Baker, 'you asked for influence. I have given you responsibility. Influence without responsibility leads to irresponsibility'. If governors have come into governing with no more than a vague notion of exercising influence, then the demands and responsibilities may come as a shock. Heads are probably less shocked in taking up headship. But a little while in the job soon proves that, without responsibility and the authority and power that it confers, it is virtually impossible to influence anything.

We believe in governing bodies. Their existence and place in the system are necessary. They embody some of our core values for education:

- It is hazardous to make education the province of professionals only, (so, this book is for heads and governors together).
- It is good to build and exercise powers through people, in this case governors, who derive their authority from acting as responsible citizens. Not to have governing bodies would weaken our democracy.
- Representativeness is a democratic reality. It works on the basis that if one person is thinking, feeling and holding a particular view, then probably others are, too.
- Team work is intrinsically better than solo operation.
- Responsibility, authority and accountability belong to the body corporate and not to individuals.
- Feelings matter.
- It is necessary to the task to be explicit about difficulties and conflicts, and the process for handling them.
- Education is about growing and developing, rather than about filling, informing, and imbuing. Life is more about cooperating than it is about competing.

The first section of this book, Chapters 1 to 5, comments on five important facets of governing: (1) governing in order to raise quality,

(2) the context of political life, (3) the language we use, (4) ownership and (5) accountability. We think these are the neglected areas for most heads and governors. The second section pulls together these important facets: (6) getting the picture together and (7) the governing body getting its act together. Chapter 8, 'Watch this space', is a footnote for your move into your new thinking world as a head or governor.

## Practical help?

'Will it be a practical book?' we were asked. A head shapes our answer very neatly: 'A set of principles', she said, 'rarely leads to one solution; it does, though, constantly inform the debate, and it rules out certain options. It serves to strengthen in a school those things that make a difference and that make one school distinct from its neighbours'. Making clear their principles and choosing the things that make a difference are what we think governing bodies need. The head went on: 'We all find it easy to lose touch with the authority of our role. We give more power than we need to gods, dragons and – dare I say – to nanny figures outside.... Under LMS not only do we have the authority to do the particular thing under discussion, but, because of the changing role of the LEA, if we don't do it, it will not be done'. By attempting to articulate and answer some of the fundamental concerns of governors and heads about the new governing body, we hope to shed some light that will enable you to make wiser decisions, based on your principles, and to achieve greater satisfaction, without nanny! If it does that, then the book will be immensely practical.

## References

8. Secretary of State, Kenneth Baker, reported in the press, at the 1989 Conference of the National Confederation of Parent Teacher Associations
9. 'A set of principles', Maggie Pringle, in *Accountability in Schools*, edited by Tyrrell Burgess, Longman, 1992

# Section 1
# An Analysis

# Chapter 1

# Quest for Quality

**In this chapter ...**

Why should we, the governors, be concerned about quality?

Why isn't the involvement of governors already working?

How can quality of practice be encouraged and enabled?

What's the actual context for making quality happen?

When, where and how do we take action on quality?

What should concern us in our quest for quality?

It is over two years since Her Majesty's Inspectors (HMI) reported on the quality of governor training. They said that the majority of governors were doing a lot of tasks well but they went on to remark that many governing bodies failed to appreciate that they had anything to do with the quality of the school: in other words it is not uncommon for governing bodies still to see themselves as tangential to the main business of the school.

> We want high standards, sound learning, diversity and choice in all our schools.... To ask the best for every child: to ask the best from every child. Excellence must be the key word in all our schools; that is what our children deserve (from the Foreword, by John Major, to Choice and Diversity, Education White Paper, 1992).

It is part of the role of a prime minister to articulate what matters in a way that ordinary people can hear. We believe that ordinary people, parents and governors, staff and heads, have heard, and do have aspirations and visions for the education of children. That is where we begin our argument that governors should be concerned with quality. We believe that governors want to do a job that is worthwhile and will only stick at it if they can be seen to make a difference.

## Why should we, the governors, be concerned about quality?

We see many good reasons why governors, as well as heads and staff, should be concerned about quality; we list them in Table 1.1.

**Table 1.1** *Why should we, the governors, go for quality?*

| | |
|---|---|
| 1 | because ... it's the real yardstick for deciding priorities |
| 2 | because ... values are the core of a child's education |
| 3 | because ... people contributing helps others to learn |
| 4 | because ... the government says so |
| 5 | because ... the government can't *secure* quality itself, it needs local contributions |
| 6 | because ... Her Majesty's Chief Inspectors (HMCI) see no point in having governing bodies if they don't go for quality |
| 7 | because ... as a fact, that's what governing bodies are doing, partly, already |
| 8 | because ... you can't separate quality out of other things |
| 9 | because ... specialists – the staff – need direction, a framework, and active partnership |
| 10 | because ... HMCI will assess the governing body's contribution |
| 11 | because ... the consequences of *not* getting involved are hazardous for the school |
| 12 | because ... it's practicable for governors to contribute |

... and because governors who only want to do a job that is worthwhile become motivated to 'go for quality'. In fact, many governing bodies are already getting there.

But perhaps you need to be convinced? Let's have a look at some of these reasons in a little more detail.

## Quality – the real yardstick for deciding priorities

The sense of overload is very real for many governors, given that a school has many facets; Figure 1.1 gives an overview. The governing body has to maintain such an overview to ensure that its planning and monitoring hold together. It is all too easy for its work to be carried out in an ad hoc manner.

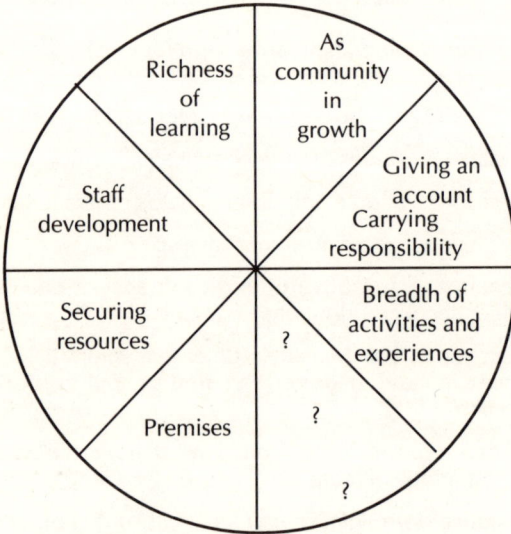

**Figure 1.1** *What do we mean by the facets of a school?*

## How do we define quality?

Debates about what constitutes quality in education will never end. We all have our views and our values which influence our definitions of quality. At its most reduced, quality becomes restricted to academic achievements and the development of basic skills but the arguments over league tables show that many parents and professionals are not content with this. The experience that a child has, the kind of person that she or he is becoming, the value that the school is adding to the child's own talent and the other influences of home and community, all matter in any definition of quality. National Curriculum attainment targets are only a beginning. Each school needs to engage with its community to arrive at its own, richer definition of quality. The governing body is the appropriate location for debate and discussion: its very composition brings together representatively the range of stake holders in a school. Before this disparate group of people can make any meaningful decisions, they must establish a common sense of purpose, a corporate view of quality.

The model of school effectiveness shown in Figure 1.2 may help you to initiate a discussion in your own governing body. What matters is that you establish your own criteria for identifying quality.

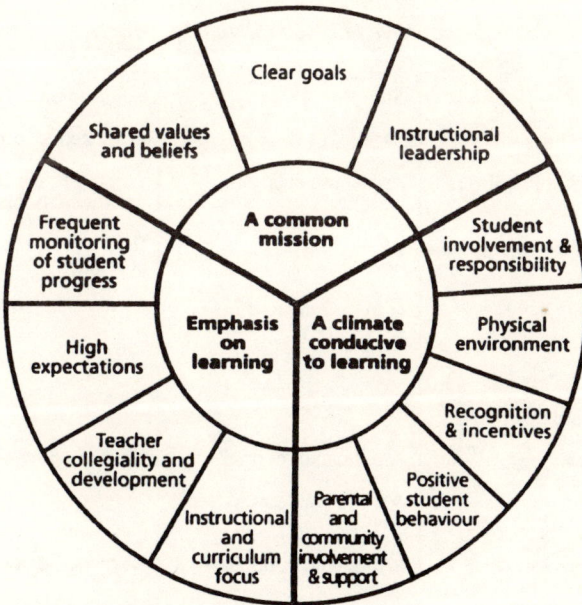

**Figure 1.2** *Halton's model of school effectiveness*

The overview of the facets of a school is what the governing body can contribute to the discussion with staff and parents. If the head and governors can make time to talk together about the quality and qualities they want, they can begin to make sense of their responsibilities and their workload. With *their* yardstick of quality they can select the development proposals that matter most for the school. They can choose, from the items swarming to be included on the agenda of their meetings, those that are urgent, or important, or both. They can more confidently let some things wait.

## Values are the core of a child's education

The 1993 Education Act requires that the school's values be encapsulated in a statement. Those who are most concerned are the parents, the head and staff, and the governors. That's not to say that the LEA and government are *un*concerned, but for them this school is not special; it is one of many. Getting agreement on the values, on quality, will take time, and the discussions and negotiation will not necessarily reach uniformity of view; the responsibilities and the perspectives of each party may be too different for that. Even their key questions may be different, as we show in Table 1.2.

**Table 1.2** *The key perspectives on school quality, 'The heart of the school'*

| Parents | Staff | Governors |
|---|---|---|
| Is my child happy? | | Is the school popular? |
| Is my child doing above average? | Are all the children learning? | Where are we in relation to other schools? |
| Is my child behaving like me? | Would I send *my* child here? | Do children keep coming? |
| | Is the school a community? | |
| Is the school clean? | | |
| Is the school normal? | | |
| Do the staff care? | Are the staff caring? | Have we got good staff? |
| Is there one member I like and trust? | Are the staff committed? | |
| Is there someone in charge? | Is the head supportive? | Have we got a good head? |
| | Is the school forward-looking? | |
| | What can we do about it? | Are the parents tranquil. . . or muttering? |
| Can we put up with it? | Can we put up with it? | What can we do about it? |

A good working agreement probably includes a recognition of the differences and of the lesser priorities, and some regular checking that the understandings are still shared. The Annual Parents' Meeting (APM) is one forum for a well-prepared discussion of values on one or two areas of school life and learning.

## People contributing helps others to learn

Children learn best when they are engaged with what they are learning, participating beyond listening, and taking some responsibility for themselves. Adults learn the same way. If the issues to do with quality in the school are just passing the governors by, then, in that part of the school, the school is not working as a learning centre. But when the governors and the head in the governing body *are* engaging with the quality issues that matter for the school, such active participation generates positive feelings for all who work there. On the same principle, Local Management of Schools (LMS) is *educationally* a good thing, quite apart from its relevance and effectiveness as a way of managing. Let's put it another way: democracy is good for people growing, democracy is good for learning.

## The government needs local contributions

There is no problem with this in itself; citizens who elect a government give it some of their authority. The problem arises when we accept unthinkingly everything that the government then says, ie when we stop using the rest of our authority as citizens to argue back.

The government *could* choose to try to do everything itself; instead it has chosen to use the model of a quango (a quasi-autonomous non-governmental organization) – in other words, the governing body – to implement its policies. The governing body has nominees and elected persons, and it autonomously co-opts more people to join it; its staff are *not* civil servants. So the government has chosen to stay out of this local field of government, hoping that the local organization which it has created will exercise the responsibilities and authority. The government has provided the necessary powers and is thus dependent upon the governing body, with the staff's skills, to secure quality.

The government *is* doing two things for itself: setting some standards of quality through the National Curriculum, and setting up some assessing and inspecting arrangements. But the middle stage of the process, making quality happen, taking and supporting action, is the governing body's task. Heads and governors have the government dependent upon them. They too are dependent, upon the staff, and they upon the parents and pupils. Unlike the government, however, governors are close enough to be able to work with the staff and parents and pupils.

## Quality is not separate from other motivating factors

The HMI survey of governor training in 1991 showed a marked reluctance on the part of governors to engage with quality. That is perhaps not surprising, because it's clear that people become governors for different kinds of reasons. Some come in with policies in mind; in that sense we see them as policybrokers, as politicians. Some come in to represent, protect and advance the interests of others, particularly parents and teachers; it's convenient to call them representatives. Yet others come in because they want to join in, be part of the school, be a general contributor; we could call them board members. The important point here is that each of these three kinds of governors will have different views and values about quality. This *ought* to lead to dynamic discussions about quality but, more often, individual governors subjugate their views to avoid disagreement.

Table 1.3 is a typical governing body agenda. It comes from the autumn meeting of the governing body to which one of us belongs. We list the bald agenda headings and the issues that were actually raised around these headings. In the third column we have used the headings from Figure 1.1 on page 14 to suggest which quality issues may have been implicit in that governing body's discussions, with its various views and values.

Unravelling the quality element of a particular item is not always easy. One could say, however, about almost any item of business, that it matters for the quality of education in the school; it's true even about a new ansaphone that frees the head to be in the classroom more. The life and learning of the school is one whole. The governors with the head just need to see it that way and then concentrate on what affects the quality, and less on making the school work. That's for the head and staff.

## The staff, as specialists, need a framework and direction

Any professional group needs to be contracted-in. The governing body is the prime contractor of the staff of a school. It is that staff's responsibility then, to galvanize their expertise, experience and energy in the creation of learning and social situations. The governing body can help teachers focus on those priorities by itself acting as interpreter of local community feelings and needs, prioritizing and monitoring objectives and arrangements at the level of the whole school, and acting as an agent of accountability to the local community. That is why Her Majesty's Chief Inspectorate (HMCI) will assess the governing body's contribution. The new inspection arrangements from September 1993 include the assessment of such evidence as:

... agenda and minutes of governors' meetings,
... meetings involving governors,
... discussion with governors involved with planning and development

[and] a statement about how effectively governors have fulfilled their responsibilities with regard to the curriculum, staffing, finance, charging; have set appropriate aims for the school; have agreed policies for special education needs, religious education and collective worship, sex education, equal opportunities, and other statutory functions. The statement should include a comment about any particular issues facing the governors. (OFSTED, 1993)

Table 1.3 *Quality issues in a meeting of the governing body*

| The agenda heading | What came up | Yes, there is a quality angle |
|---|---|---|
| Introduction of new clerk | | |
| Election of chair | | Governing body development? |
| Election of vice-chair | | |
| Minutes | | |
| Matters arising | * Co-options<br>* Membership of working party for the governors' annual report and the APM<br>* Tabling of the development plan | Giving an account and carrying responsibility?<br><br>Richness of learning? |
| The head's report | * Membership of the working party on pay policy<br>* New clubs at dinner time?<br><br>* Visit to the outdoor centre?<br>* The evening for individual parents with teachers? | Staff development?<br><br>Width of activities and experience?<br><br>Giving account and growth as a community? |
| Any other business | * Getting the parents' view on discipline<br>* New telephone and typewriter for the general office<br>* Membership of the working party on monitoring the budget<br>* Hazards from cars collecting pupils<br>* Hazards from dogs at gate | |

To forestall criticism from the visiting inspectors is not, in our view, a good enough reason on its own for governors to go for quality. But the quotations make clear that the Office for Standards in Education (OFSTED) sees some cause-and-effect connections between the work of the governing body and the quality of the school that is being inspected.

## The consequences of *not* getting involved are hazardous for the school

We've already suggested that excluding anyone from the life of the school, or anyone excluding themselves, runs counter to the principle and process of a learning centre. If the head wants and manages to exclude the governors from quality issues, that will have limiting effects on the staff and on the parents. The head is neglecting an opportunity and is failing to shape a demand force which could be powerful and locally persuasive on behalf of the school. 'If governors aren't part of the resources, then they're just part of the hassle', said one head. 'A potential millstone', is how others have seen it. Not sharing responsibility leaves the head isolated if there's any vulnerability, blame, or serious mistake to carry. In contrast to the parent body, the governing body *is* a partner with whom the head can work on level terms. The hazards to the school, if the head monopolizes, seem obvious.

We have deliberately put this section in rather blunt terms. We have deliberately referred to the possible monopoly of power. Yes, we know about governors' inertia stemming from overload, their relative incompetence due to their newness. We know also, from the evidence of a survey commissioned by the National Association of Head Teachers, that the great majority of heads expect 'to manage the Governing Body' that for the great majority the governing body is not a significant contributor. That evidence sits squarely with our own experience of heads.

We believe, for all the reasons we have set out above, that the governors' key contribution is to quality issues. We believe that this does not pose a risk for most heads; the risks from drawing rigid lines which say 'thus far, no further', are greater. We believe that heads and governors are capable of partnership, even when defined in the following rather demanding terms:

[A partnership is]

a working relationship that is characterised by
a shared sense of purpose
mutual respect
and the willingness to negotiate

This implies a sharing of
information
responsibility
skills
decision-making and
accountability.

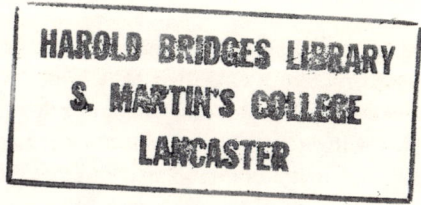

(Gillian Pugh)

The head is in a unique position to foster or to block the partnership. With this challenge, we rest our case. We hope you are convinced that the quest for quality is worthwhile and necessarily involves governors.

## Why isn't the involvement of governors already working?

### Gatekeeping

The head and chair as a pair are likely to function as gatekeepers for the governing body, managing the flow of business in an orderly way. The head has the day-to-day responsibility for the whole school and the knowledge, skills and experience in teaching, leading and managing; the head has paid time. The governing body probably has no independent location. It has very limited alternative sources of information about education and the school, and has only the barest minimum of staff time available to it. The governing body rightly sees the head as its chief executive officer and chooses the head to be the gatekeeper for the school – and the gatekeeper for the school is necessarily involved in gatekeeping for the governing body. But who establishes the rules and directions for your gatekeepers? Why aren't the governors coming inside the gate? Why aren't the governors effective inside the gate? Why is the working partnership difficult to achieve? We need to look for the answers in the ways people generally behave and interpret each other's behaviour:

I'm determined. You're stubborn. They're pig-headed.

We probably all use the best terms for ourselves, the nice terms for those around us, the derogatory terms for 'them' – and the actual behaviour is probably the same! It's easiest to put blame 'out there'. So, from the point of view of governors who want involvement and who find that the involvement is not working, the reasons we will see most easily are likely to be the reasons 'out there'. We governors will be likely to underestimate the other real reasons that may lie nearer to us, and we may even ignore or deny the reasons rooted in ourselves. The importance of each factor or combination of factors will vary from school to school. Each governing body must measure for itself how significant a particular block is. Our 'block-o-meter' in Table 1.4 is not comprehensive but could be used as a neutral tool to start you and your fellow-governors, including the head, thinking. Let's look a little more closely at some of the possible blocking factors.

**Table 1.4** *Block-o-meter*

| THINGS | |
| --- | --- |
| OUT | Head keeps key issues and fobs-off governors |
| THERE | Head clutters the agenda with trivia |
| THE | Struggles in the past demoralize the governors |
| | The government is unclear about ... government ... managing ... executing |
| CLIMATE | People in education are unclear about governing |
| | Managing is not high-prestige |
| AROUND | Committees tend to go for the easily-handled items |
| | Corporate working in a committee is difficult |
| US | 'Governors must not spend money on their development' |
| | It's hard to see the wood for the trees |
| AMONG | Governors don't want to be blamed by government or anyone else |
| THE | Governors are diffident in speaking about quality |
| GOVERNORS | Governors feel powerless and don't exercise their muscles or their minds |
| | 'We are only volunteers' |
| | 'We are only amateurs' |

Let's look a little more closely at some of the possible blocking factors.

## Governors feel powerless. Governors are diffident

Quite properly, governors are diffident because improving the educational experience of children is not simple, not like a consumer having to deal with faulty tins of fruit or faulty shoes. Sadly, we have a culture that puts pragmatism – 'can do' – higher than matters of principle like democratic participation. Unfortunately, our respect for the professional's skills often tips over into undue deference to the professional's views; the amateur is undervalued; we lose the parity of esteem for different contributions that characterizes a team approach. Understandably but regrettably, the teacher professionals had defined quality from their perspective alone, to do with teaching approaches and the learning process, the academic life rather than the wider life of the school; others, including governors, have been remiss in not making their contribution to widen the definition of quality; and governors now live with the consequences of that biased or skewed definition. Particularly for the professional teachers, the National Curriculum's first fling was too prescriptive, not allowing scope enough for professional discretion so that teachers feel themselves to be hacks, merely technicians. That makes it difficult for governors to get into discussion on curriculum and quality issues.

## It's hard to see the wood for the trees

It's hard, we suggest, in two ways at once. (1) At a certain distance you can see the trees, but not the wood as a whole; the governor can see some aspects of the school, but cannot see the school as whole. (2) At a certain distance you can see trees, but not the grain of any individual tree; the governor cannot see what learning is like in the classroom; can see an argument for more science equipment, but cannot see the life of learning science in the classroom. The consequence is that the governor may have to change positions and angles of vision, frequently. If in doubt, the governing body should stay where it can see the wood for the trees – the school as a whole. That overview is the governing body's particular responsibility and one they should stick to, rather than being, as some intrusive governors are felt to be by their long-suffering heads – wood-peckers or nit pickers!

## 'Governors must not spend money on their development'

This attitude seems widespread. The government reinforces this attitude by the money it allocates for governor training and development: £15m for

300,000 governors is equivalent to £50 per governor per year or £200 in a four-year period of office. But the attitude comes from deeper roots than the government's token sums. Our culture doesn't rate meetings much, doesn't value the process of hearing different perspectives, nor the negotiation of a working agreement, all of which are crucial to a governing body's effective working. The role of governors is left too inarticulate and confused. So governors are sometimes unconvinced of their potential effectiveness. And government, parents, most heads and staff are by no means consistently encouraging, enthusiastic and supportive; some heads, however, are the best midwives and educators of the governing body. In such circumstances it's hardly surprising that few governing bodies at present plan their own development alongside that of the staff.

## Committees tend to go for the easily-handled items

How true is this statement of your governing body at work? Such practice does enable the task to be comfortably manageable, it makes a low-level decision and some modest satisfaction quite possible, and it minimizes the passion of arguments over values and clashes over important alternative futures. When it looked at the working of committees, the Audit Commission found a picture that will not surprise governors:

- up to 50 per cent of the time was spent *receiving information*, (*not* the information necessary for a particular decision)
- anything from 45 per cent to 65 per cent of the time was involved with *operational matters*, (what are the staff for?)
- only 10 per cent to 25 per cent of the time was to do with *performance-review and policy-making*, (which is the area of a committee's potentially unique and vital contribution).

## For discussion

Some of the other blocking factors which we suggested may be self-explanatory. You may like to discuss any that seem relevant to your situation.

- Managing is not high prestige
- People in education are unclear about governing
- Struggles in the past demoralize the governors
- The head clutters the agenda with trivia
- The head keeps the key issues, eg the development plan, obscured by detail, and fobs off the governors with monitoring the temperature of the radiators (yes, it really happened!)

For heads and governors the school is a large and complex thing, which is one of the reasons why it can't be left to the head alone to plan, direct, manage, operate and monitor the school. Glance again at Figure 1.2 to remind yourselves of the many characteristics of effective schools.

No wonder that, in the few years since the 1980 and 1986 Education Acts regalvanized governing bodies, they haven't yet made a total success of their contribution.

## The government is unclear

We noted in an earlier section that the government has chosen to stay out of this local field of government, hoping that the local organization which it has created will exercise the necessary responsibilities and the authority. That general freedom which has been left to the governing body seems to us to allow it to govern. The key issue is freedom. There is always the temptation for government to encroach on its own proclaimed principle of freedom. The ever-growing list of information which the government *requires* the governing body to include in the governors' annual report seems to us to relegate the governing body to being an executive publishing agency. The government has fallen for the temptation of trying to manage the education service and into the trap of having to manage the grant-maintained (GM) sector. Hence the tidal wave of Acts and the permanent flood of Regulations to fill in the details, hence this protest from a head in North Yorkshire who had attended a 48-hour conference with 'the managers' from the Department for Education (DfE):

### A Cry from the Heart

Sir – The DfE's conference this year was entitled 'Drawing the Line', and officials convened a hundred folk, HMI, Chairs of Governing Bodies, Governor Trainers, and a few head-teachers, to spend three days together drawing that line.

They meant the line between Governing Bodies with their Chairs and Head-teachers with their management teams. We worked hard and long but came to think that there was no such line; nor should there be. Instead we all cried out loud and clear against the line that the Secretary of State and his Ministers have drawn around themselves and DfE officials; a very few of the officials, we came to suspect.

The sense of grievance, the sheer frustration, the tangible anger felt against the extremism of the Government's educational policies and its utter unwillingness to consult or to listen, were tangible; so too was the concern that the strength of feeling would never penetrate that self-imposed boundary line. But maybe one of the triumvirate or the mandarins reads your letters page?

Head-teacher

PS: One thing the conference did prove. The DfE and its ministers have succeeded in uniting the education service, but sadly it is united against them.

By cutting out the LEA as the middle manager the government has laid open to confusion its own role and that of the governing body. Both are tempted to stray into levels of detail that are inappropriate from their position in the system. No wonder heads and governors haven't got it together yet. Our next three sections in this chapter offer you ways through the tidal waves and floods.

## How can quality of practice be encouraged and enabled?

Three things seem to us to be needed for the governing body's work with others to extend and to raise the quality experienced in the school:

- Think wide. Target local and national government.
- Work at partnership with *the staff*, *the parent body* and the *parents individually*. Most governing bodies have hardly begun to think seriously in these areas, despite well-intentioned words.
- Set out to negotiate with other partners. Know what you want and be prepared to listen to their wants.

### Think wide

Target local and national government. Go to government. Feed your views to the DfE and to OFSTED, to the Schools Curriculum and Assessment Authority (SCAA) and to the Funding Agency for Schools (FAS). Nominate a good letter-writing governor, (1) to receive material, to get others to think, and to make sure responses are made, and (2) to send some comment, report, or request each term relevant to the priorities in this school. Go to the LEA. The role of the LEA is changing, but its decisions, eg on funding, have a huge impact on your school, even if it is grant-maintained. Make your relationship with its officers and members a particular one that they will remember among the 100 or 500 other schools in their concern. Put your MP on your mailing list. Give him/her a sense of the life, triumphs and difficulties in your school. No need to ask them for anything, yet. Educate them. The ground-swell of MPs' opinion does emerge as party policy for education.

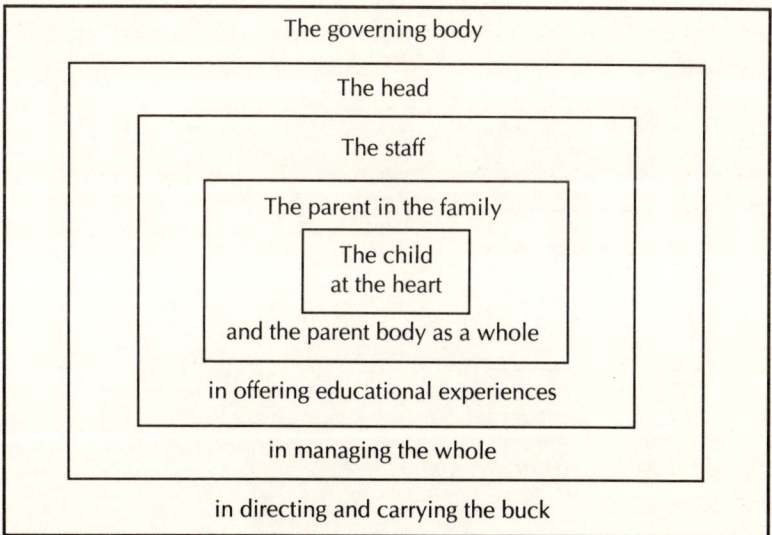

**Figure 1.3** *What do we mean by 'the school'?*

## Work at the partnership with staff and the parents

One important question which most governing bodies and staff don't discuss is, what do we mean when we say 'the school'? Who do we include? Who do we exclude? Because we do not think through and agree our definitions, there is often unnecessary confusion. We offer a view in Figure 1.3.

We hope that this diagram widens the horizons. We don't find that most heads, staff and governors put parents in or near the middle of the school. Certainly, heads, staff and governors give a lot of thought and time to contact with individual parents. But we find that it's more of a reaching out to individual parents, and more of a bringing-in of all the individual parents for a particular purpose or occasion. The picture in the minds of heads, staff and governors is probably more like Figure 1.4, with the school reaching out and the parent coming in. The child, of course, goes along both arrows, in and out, every day. We think that the concept of the parent body and of individual parents actually *existing inside* the picture of the school is more educational, more participative, more constructive in principle. No easier to make it work, however. And, of course, it doesn't mean parents spending all their days in the school. From the individual parent's and the parent body's points of view it's the difference between feeling oneself an insider or an outsider.

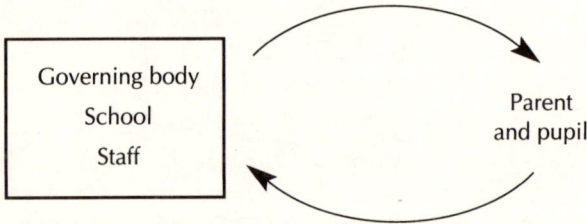

**Figure 1.4** *Reaching out and going in*

We believe that school quality rises most readily when the parent body is active for quality. There's a lot of evidence to support this, ranging from reading partnership schemes and the organizing of clubs, to the 20 per cent of time that APMs really get into a discussion. We suggest that it may be easier for the governing body with the staff to develop that responsible co-ownership with parents, easier than leaving it to busy staff on their own.

The government has set standards and indicators of quality and is introducing new arrangements to inspect and to report nationally on quality. As part of that, all parents will be invited to give their views and to meet the inspectors before they begin work, as shown in Table 1.5.

A summary version of the inspectors' report will go to the individual parents from the governing body, along with a copy of the governing body's action plan which it produces in the light of the report.

**Table 1.5** *Inspectors' questions to all parents about school quality*

| | |
|---|---|
| The school: | values and attitudes taught |
| | standards of behaviour set |
| | efforts to secure attendance |
| | encouragement to parents to play an active part in the life of the school |
| | sense of welcome |
| | range of subjects taught |
| | provision for pupils with special needs |
| | |
| The child: | satisfaction with child's standard of work |
| | information about child's work |
| | help and guidance given to the child |
| | homework |
| | happiness of child in school. |

## Set out to negotiate

Receiving the paper-work may not be enough. It may be that the parent body and the individual parents will need to be involved in the shaping of the action plan, will need to agree to take action as parents, rather than simply to receive what's going on as 'customers'. All the evidence suggests that parental influence has a greater effect on the child than anything else. Kenneth Baker, who was the Secretary of State at the time of the Education Reform Act, never doubted the importance of parents to education reform:

### A joint responsibility

Whatever reforms I am able to bring to the education system, they will not produce the results we all want to see unless children are able to benefit from schooling. Whether they are able to benefit is something determined more by the attitudes of the home than by the attitudes of the school. We rightly look to teachers to teach, but it is the joint responsibility of parents and teachers to educate.

We often talk about the partnership in education. Of course parents have a right to expect schools to provide good education, and that is why we are undertaking radical reforms of the education system. But perhaps we lay insufficient stress on the responsibilities of parents in that partnership. Teaching is a difficult enough task made even more difficult when parents don't take their responsibilities seriously enough.

We recognize that there are real tensions around the government's view that the parents know best. Yes, for some things, but not for everything. That's why the Education Act 1944 made parents responsible for the education of their children. But the Education Act 1944 required LEAs to secure 'that efficient education should be available to meet the needs of the population of their area'. 'Available' is the key word. Simply to say that parents know best, implying 'about everything', distorts the reality.

The partnership for quality which the governing body is best placed to foster is of parents, staff and governors *active* together, as in Figure 1.5.

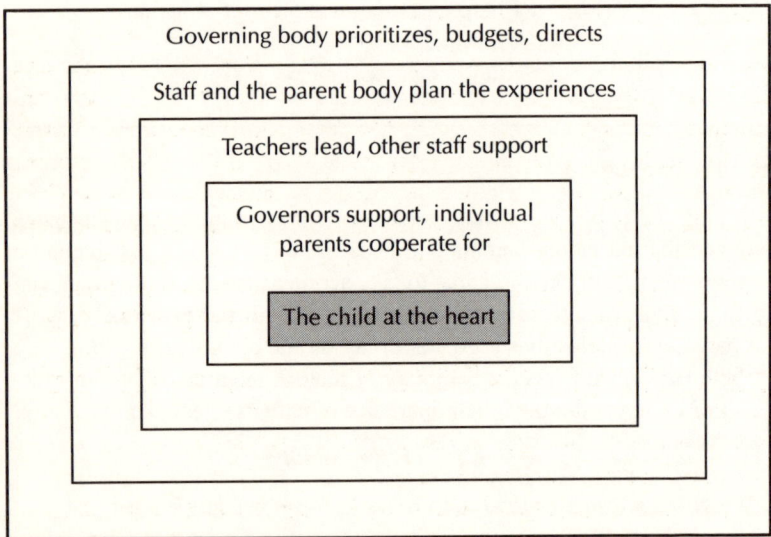

**Figure 1.5** *School contributions to the partnership of learning*

As with most things, there are words and issues that tempt people into a fuller partnership. A lot depends on the questions which the governing body puts to its prospective partners. Table 1.6 summarizes three broad approaches, which fit within three different *kinds* of partnership.

**Table 1.6** *The governing body's approaches for different partnerships*

> Would you please *endorse* our proposals?
>
> Can we *discuss your views* on these ideas of ours?
>
> Can we *negotiate* together what might be best?

You will get much more response, thought, commitment and actual help from your partners the more issues you can find for which you can use the negotiating approach. In education no one can afford the limitations of being *sole* owners

## Negotiating with the government

The government has taken over virtually all funding, severely restricting the LEAs' capacity to raise funds. The government has always been responsible for teacher training. The National Curriculum and its assessment are under the government's control. The government has practically eliminated the LEAs' capacity to offer advice and support, and is relying on the market to generate alternatives. The government is pushing information, but is leaving a network of contacts to be made by personal effort.

So what does the governing body need to negotiate with government in order that the governing body can secure quality in the school? In Table 1.7

**Table 1.7.** *The context the governing body needs to work with quality*

> Sufficient funds, calculated on what needs doing locally.
>
> A market supply of well-trained staff.
>
> A national definition of standards, for use alongside local standards.
>
> A dependable source or centre of advisers, problem-solvers and developers.
>
> Information and a network of contacts for the governing body to use in making comparisons.
>
> The principle that the governing body has some subsidiary authority in representing the local community. The governing body is not dependent on statutory authority alone.
>
> The principle that the government will work with a formal consultative body that represents governing bodies in negotiations about the running of the education service.

we suggest five specific resources and two general principles which show what we mean by the advice to think wide, work at partnership, and set out to negotiate.

## When, where and how do we take action on quality?

We suggest that there are four focuses on which a governing body can practicably work, as its basis for action on quality:

- A development committee or working group.
- A development plan or paper.
- The Annual Parents' Meeting.
- Self-appraisal of the governing body itself.

### A development committee

Most governing bodies use standing committees or ad hoc working groups for some of their work; a few try to handle all the work together. Many governing bodies have separate finance, personnel and curriculum committees. We suggest that separating curriculum from the rest of school life narrows the development that may be appropriate. We find that putting finance on its own may make it just a monitoring-of-expenditure committee, or alternatively – and equally mistakenly – a committee whose grasp of finances gives it power over the whole range of policy-making issues. There seems more sense in creating a single development committee or working group. Its scope would be all development needs. Its task would be to prioritize them, and to plan the financing of them in the context of the school's current work. (The monitoring of expenditure could be by one or two governors with the head and bursar, reporting to the governing body at every meeting.) Clearly, such an integrated development committee would be the key committee. It would need to share its thinking with all governors and all staff, and in part with parents.

### A development plan or paper

Regardless of what development information the LEA or the DfE want from the school, the governing body needs to have some paper or plan that shows what the school plans to emphasize in the coming year in terms of its growth.

## The Annual Parents' Meeting

The APM is potentially an important event which gives,

> the whole parent body the opportunity to become more closely involved
> in the life of the school, to mutual benefit. Governors will wish to create
> an atmosphere in which parents can freely express their views, and both
> governors and parents can engage in full and business-like discussions
> (DES Circular 8/86, December 1986).

The 20 per cent of life in APMs seems to happen when governors and the
head want to listen (not just to comments about the governors' annual
report), and because governors and the head discover the issues in the
school's life and learning on which the parents want to have a say. The
developing and increasing quality of the school is something which matters
to parents. Discussing some aspects of what may go into the development
plan is clearly something where their views will help.

## Self-appraisal of the governing body

The governing body should make time to tackle the questions, 'How much
do we think we are contributing to the effectiveness of the school?' and,
'How well do we think we are doing as a team?' The answers may be as
simple as Table 1.8.

**Table 1.8** *Self-appraisal answers for the governing body*

| How much do we think we are contributing to the school? | | | |
|---|---|---|---|
| Little? | Something? | Useful enough? | A great deal? |
| How well do we think we are doing as a team? | | | |
| Poor? | Fair? | Well enough? | Very well? |

It will help the governing body to set targets for its contributions in the coming year, a sub-plan for the development plan. It will help the governing
body decide what things it needs to *advise* on, what *planning and policy
making* is timely, what needs *coordinating* and which tensions need *mediating*, where the need is to *promote* and to *support*, and what particularly to
*monitor*. It will reveal where the governing body needs to *improve its own
working*.

### Inspecting the school?

The four-yearly inspection as part of the government's national arrangements is a new additional process. While governors are not inspectors, the governing body needs *each year* to review the strengths, weaknesses, opportunities and threats encountered by the school; it needs to know of progress made and limitations met. The development plan is a consequence of that monitoring, but there is a need for the governing body and staff to establish some arrangements for monitoring as a continual part of school life and learning.

Monitoring expenditure is easy. Monitoring quality is difficult. Should governors themselves be among the observers doing the monitoring and reporting? Will a variety of angles, the governors' as well as the head's and the staff's, lessen the natural self-interest of each party? What is the staff's view of the governing body's good sense and integrity? Do all accept that forming judgements is valid, even if occasionally distressing? For us the questions are finely balanced. They need open discussion by the governing body and staff. So long as the governing body gets as clear and fair an annual monitoring report as that of the four-yearly inspectors, we don't think it matters who does the work.

## What's the actual content for our concerns with quality?

### Everything

The answer to the question is *everything but not equally,* and *you*, the governing body, must develop and use your own yardstick. To be most helpful, it will be based on your agreement about values. The governing body has to have an overview that holds all the pieces in the frame of its planning and monitoring. Decide which few dimensions of the school, held alongside each other, will give you an overview.

Look back at Figure 1.1 (page 14) on the facets of the school, and at Figure 1.2, the Halton model of the characteristics of effective schools. Get one or two governors to look at OFSTED's *Handbook for the Inspection of Schools* – a copy was sent to the school – to see if that could help the governing body in putting together its overview. As a more selective range of items, Table 1.5 (page 29) listed the inspectors' questions to all parents about school quality.

The overview given in the 1988 Education Reform Act is one that you should think about:

*1988 Education Reform Act*
*The curriculum to be secured by the governing body or head*

A curriculum which:

- promotes the spiritual, moral, cultural, mental and physical development of children,
- prepares pupils for the opportunities, responsibilities, and experiences of adult life.

It needs to be balanced — an agreed amount of time for each subject so that all curriculum areas are covered,

broadly-based — a wide range of knowledge, understanding, and skills,

relevant — to the pupils' experience and future life,

and differentiated — matched to pupils' needs, aptitudes, and abilities.

What we're saying is that every aspect of school life and learning has some impact upon the quality of experience of the child. The governing body has to use its own yardstick to decide which needs in any one year are the most important and the most urgent, for instance, which things need advising on, for their relevance to quality? Planning and deciding, for their relevance to quality? Co-ordinating and mediating, for their relevance to quality? Promoting and supporting, for their relevance to quality? Monitoring, for their relevance to quality? Which things, in the governing body's own ways of working, need revising in order to make a greater contribution to the school? It all depends on how your governing body sees it, how you interpret it.

Our final offering in this section is from an anonymous correspondent in Column One of the weekly journal *Education*:

### The Clerke's Tale?

A cynical guide for new school governors to this term's Governing Body agenda.

1. Apologies for absence (*EastEnders* is preferable)
2. Election of Chair (The Power Game resolved)
3. Election of Vice-Chair (Consolation Prize)
4. Appointment of Co-opted Governors (Those who have been conned)
5. Minutes of last meeting (Ramblings on the Un-read)

6. Head's Report (A sometimes subtle effort at self-congratulation)
7. Draft Policy Statement on Special Needs (The Conscience Clause)
8. Reports from Committees/Working Parties (Making a meal of the unnecessary)
9. School Budget Items (The partially-sighted leading the blind)
10. Parental Complaints Procedure (The Moaning Minnies' Charter)
11. Governor Training (Even *Neighbours* is preferable)
12. Annual Meeting for Parents: 5 November (Damp Squib)
13. Head-teacher Appraisal (Head-to-Head back-scratching)
14. School Entrances: Car Parking (Failed diplomacy with selfish parents)
15. LEA/Governing Body Partnership (Wish we knew what LEA does)
16. Pupil Exclusions (Abandoning the unfortunate)
17. Log Book (Head's autobiography)
18. Any Other Business (Depends on the time)

The success of a Governing Body is dependent entirely on the governors; it is not a Head's consultative meeting, nor an LEA Committee, nor an opportunity to promote party political, nor indeed Government, propaganda.

'The success of a Governing Body is dependent entirely on the governor'; we agree. We'd add: 'Don't forget that the head is probably the only person in full-time governing'.

# References

12. *The Quality of Training and Support for Governors in Schools and Colleges*, HMI, DES 1992
12. *Choice and Diversity*, DfE and Welsh Office, 1992
19. *Handbook for the Inspection of Schools*, Office for Standards in Education, HMSO, 1993
20. NAHT research survey, Birmingham University, 1992
20. 'Partnership', Gillian Pugh in *Parental Involvement*, edited by Wolfendale, Cassell, 1989
24. Audit Commission, *We Can't Go On Meeting Like This*, 1990
25. 'A cry from the heart', Roger Haslam, *Education*, December 1992
29. 'A joint responsibility', Secretary of State Kenneth Baker, The Churchill Lecture, 1987 quoted in *The Home School Contract of Partnership*, NAHT, 1988
36. 'The Clerke's Tale?', Tacoma Bridge, *Education*, October 1992

# Chapter 2

# Politics, Philosophy and Economics Shaping Education

*In this chapter ...*

Can't we all keep politics out of education?
What's the energy that's fuelling the government's moves?
What are the energies and driving forces at school level?
What are the driving forces elsewhere in the system?
The authors' particular perspective on the changes.

Politics is about devising policies for the conduct of areas of public life. Differences in politics are expressions of different philosophies, values and beliefs. The conclusions arrived at will not only be determined by our understandings and beliefs but by what we can afford – the economics.

Governing bodies are experiencing at the local level all the tensions of a national government. When resources are in short supply, we simply cannot have all that we might want. It seems to us that the government model for arguing about priorities is not always a good one. Choices are put to us in terms of right and wrong, or better and not so good, when the real truth is that for some things, such as nursery education, we may have to wait because there simply is no money to make it happen now. Honesty about such matters may bring disappointment but it also brings a certain amount of relief that we know the score. Governing bodies would do well to come clean in their politics and their economics with each other and with the rest of the school.

There is sometimes a risk that all of us in education are so preoccupied and overwhelmed by the individual waves that are breaking on our shore that we don't look out to see what the tide is doing!

In this chapter we want to offer some comments that may help you to stand back from the waves breaking at your feet, to understand and to say, 'Oh, I see what the general drift may be!' If, from the same facts that we use, you make a different interpretation, that's fine. It's the standing back and reading the signs that matters.

## Can't we all keep politics out of education?

We think that the answer to this question is and should be 'No'. Education is an area of public life. The policies which determine it are framed in the national and local political arenas, and becoming governors and heads means that we enter that arena. One reason why the contribution of governing bodies is so much less effective than governors and heads want it to be is because they shy away from the cut and thrust of argument. They're all trying to play it by some given book of rules despite the occasional protests made at conferences. They're not behaving politically enough! There are too few arguments about the values of education for the governing of schools to be in a healthy state. Arguing is part of the politics of governing and of education.

What is it about the word 'politics' that makes people want to dissociate education from it? Is it that people's reactions include a sense of sordid, self-interested, and cheap arguments and gibes? Do people become uneasy when a discussion, which might be a fair contest of values and ideas, degenerates into destructive criticism and a slanging-match? If so, part of the responsibility must lie with elected politicians at the national and local levels. For them the potential of debate and discussion can become side-lined by the wish simply to exercise a mandate to govern for a period. Once the mandate is won, there's no real negotiation with the opposition groups, nor any real persuasion of the public. White Papers, council plans and media manipulation take the place of engaged discussion and debate. No wonder, then, that heads and governors say, 'That's no way to run this school community. We don't want *those* politics'. But politics does not have to mean that.

## Politics as a constructive process

We see politics as being the process of arguing over what matters most and over how to develop the best. Table 2.1 summarizes our argument for bringing politics into education.

**Table 2.1** *Why politics is in education*

---

- Politics is the negotiating and expressing of a set of values.
- Politics is about the exercise of responsibilities.
- Education involves values and develops responsibility.
- Therefore, politics in education is necessary ... inevitable ... desirable.

---

If education matters enough, then it's worth arguing over. And that's the process of politics. Values do not emerge from nowhere; they stem from an understanding of the way things are and a vision of the way things are to be – that's the underlying *philosophy*. The *economics* take account of the driving forces, the demanding needs, the priorities, the value for money. If all of us as governors and heads tried to exclude political values from education, we'd be excluding ourselves as citizens from one of the essential strands of education in the public domain; we'd be denying our responsibility for education; we'd be saying, 'It's their business'; we'd be limiting ourselves to putting up with what is being decided for us without question.

Some governors and some heads want to play by a book of rules. They ask questions, but without making their own answers, so that others have to write more and more regulations to clarify every nuance of every question. (Fairly clearly, the others who have to write include the DfE, OFSTED, the new Schools Curriculum and Assessment Authority, the Audit Commission and the FAS.) There are plenty of people outside the governing body willing to give answers for business that, in the end, is not on their plate. Why do we give them that power by asking in the first place? Why not work it out for ourselves, governors and heads together?

## Politics of obedience, politics of choice

As we see it, governing bodies have a simple choice. Do they settle only for governing on the principles of obedience – 'we should' – or on the principles of 'we choose'. Table 2.2 sketches the crucial distinctions.

**Table 2.2** *The choice of 'we should' or 'we choose'*

| If the principle chosen by the governing body is . . . | We should | We choose |
|---|---|---|
| then the attitude behind the choice is . . . | because the authority says so | because we have autonomous authority |
| the consequences are that . . . | the governing body is directed by others<br><br>it is only an agent | the governing body will direct some of its actions at its own discretion |
| the governing body's options are . . . | to submit or rebel | to agree or disagree |

Playing it by the book means that the 'we should' set of principles domi-nates. The consequences may well include resentment and frustration, resistance and a sense of guilt. Behaving politically means the 'we choose' set of principles are the counterbalance to government directives. It means thinking about the levers of power and who can legitimately use them. Politics comes into the agreeing and the disagreeing, accepting the conse-quences and taking responsibility for a local decision. It's more difficult! We believe it's also more human, more educational, more responsible.

In the next sections of this chapter we offer what we see as the general drift and our judgements and views on what we see, based on our values. We note briefly how far what we see is justified or reasonable, unfair or exaggerated, whether it presents an opportunity or a threat for what we see as the good governing of schools. We take it for granted that there is nearly always good and bad lying side-by-side in educational practice and in plans for education.

> Life is a rose garden.
> The petals wilt
> and the thorns remain.

# What's the energy that's fuelling the government's moves?

It is entirely reasonable and necessary for governments to have an agenda. For the governments elected in the 1980s and 1990s the agenda for education has seemed simple:

- Dealing with 'them' – any sub-group perceived to be a potential obstacle to the individual citizen's growth.
- Programmes that delegate, but also harness and control, tame and domesticate.
- The notion of individual standards.

## Dealing with 'them'

In education, this agenda has focused on dealing with teachers and the teacher associations, and 'lunatic' LEAs by a process of local management and the creation of the National Curriculum.

### Putting the teachers in their place

The government has tapped into the weight and energy of parents' discomfort with the 'distance' of teachers, their professional exclusiveness. A sharp illustration was given in the National Consumer Council's 1986 report on home and school links; it was called The Missing Links:

> *The missing links between home and school*
>
> Although schools, parents, and LEAs are trying harder to communicate with each other ... there is still an enormous gap between parents' and teachers' perceptions of what is happening.... What teachers say about home-school links does not always tally with what the parent perceives.

The distance is still there in 1993, as Sylvia West, a secondary head, noted with regret: 'Schools remain remote and mysterious places to many parents'. At the same time the government continues to recognize and to draw on the apparently insatiable desire of parents for 'higher standards', the phrase being used to justify every proposal.

And that's reasonable. Higher standards has to be *the* principal goal, for any venture in education. But the government, through highlighting limitations and difficulties in education, has put the teachers at the receiving end of most of the criticism and blame; in Summer 1992 the Secretary of State was even critical of improved GCSE results. The collective energy of teachers in their trade unions and professional associations is scorned twice over. They are scorned for being in unions at all – generally defeated and driven from the field – and for being inadequate professionally. 'The teacher-unions are so unrepresentative of the average good teacher that John Patten could more profitably spend his time chatting to animals at London Zoo.' Thus wrote a columnist of the *Evening Standard*. The government has also tapped the frustration and energies about Britain's faltering economic health. The government believes and comments that the education service has not delivered the workforce the country needs. This unduly and therefore unfairly shifts to the teachers and other professionals in the education service blame which is perhaps more reasonably due to industry and business.

Groups other than teachers have also been presented as a threat. John Major at the Conservative Party Conference in 1992 looked forward to 'another colossal row with the education establishment'. The 1992 White Paper derided 'educational theorists' and 'administrators' as knowing little about the needs of children. Chief education officers have been called to account by the DfE for their presentations of the pros and cons of grant maintained status, and parents have been described as neanderthal when disagreeing with the Secretary of State.

Up to now governing bodies have been regarded generally as a good thing, but we wonder whether the government will ever get to the stage of not trusting governing bodies. That might be a possibility if they become organized into a democratically representative forum with a voice. The volunteer members of Training and Enterprise Councils are already so organized. The governing bodies of further education colleges are developing a forum. Can the government live with, and not obliterate, dissenting and uncomfortable views? The only significant deflections so far from its chosen path have come (1) from the widespread protest at the proposal in 1991 to prevent teachers standing as parent-governors; (2) from the House of Lords' insistence in 1992 that the appointment of school inspector teams should be overseen by OFSTED; and (3) from the refusals of parents, teachers and governing bodies in 1993 to go along with unrealistic timetables for testing and assessment.

## Knocking the LEAs out of their place

The biggest 'them', seen and presented as a threat to the government's vision for education, has been the LEAs. They had discretion in local educational policy-making, local politicians with effective responsibility, and the capacity to raise money for educational purposes. In the government's determination to minimize spending and to centralize authority, the LEAs have been relegated to a lesser, administrative, role. The proportion of LEA funding under government control has gone up from 40 to 80 per cent. The LEAs have been presented as bad objects. Table 2.3 summarizes the allegations made against them.

**Table 2.3** *LEAs as seen by central government*

- Bureaucratic

- Unnecessarily expensive

- Out of touch with local people

- Restrictive of educational freedoms

- Sometimes devious in their allocation of funds.

There was some truth in all of this. But the government has set up more centralized bureaucratic arrangements, with a DfE doubled in size and cost, with no local access points for people, and a mass of legislative regulations – which now shapes the agenda of each governing body. (The 1993 Act requires an annual discussion about moving or not moving to grant-maintained status).

The government has drawn on the fear of state socialism and left-wing LEAs. At the same time it has established a nationalized education system, with what it terms 'self-governing state schools', with a single funding source. Even the churches are to be tempted to reduce their stake, as the White Paper suggested: 'The governors of ex-voluntary-aided GM schools are not liable for a 15% contribution to capital and external repair costs.... Voluntary schools have a lot to gain by becoming grant-maintained'.

The Secretary of State sees the end of a national system, locally administered. What the 1992 White Paper offered, in its opening quotation from John Ruskin, was, 'training schools for youth, established at Government cost, and under Government discipline'. The extreme left of any socialist party might have wanted no more. But do heads and governors welcome 'training' rather than 'education'? We, the authors, don't value the dependence of the new arrangements on the goodwill of the Secretary of State for the day.

There is no question that most schools favour the flexibility offered to them by local management schemes (LMS) and about a thousand schools, at the time of writing, have decided to remove themselves from the local systems operated by LEAs. But there are some disadvantages emerging, such as the loss of library services, relied upon by smaller schools, because larger schools have decided to do without or go elsewhere. The full diseconomies of smaller units have not been costed or fully felt yet. The periodical *Managing Schools Today* reported in 1992 that, 'The replacement of the Inner London Education Authority by a large number of small authorities is now estimated to be costing in extra bureaucracy some £500 million a year more than the cost of running ILEA'. No one has yet reported on the administrative and necessary bureaucratic costs in schools with their local management, in comparison with the costs of the same tasks when carried out by LEAs. The consultants Coopers and Lybrand in their DES-commissioned report on LMS in 1987, expected that 'administration by more would cost more'. Meanwhile, as Dr Mike Kelly of Manchester Metropolitan University puts it, 'Governing bodies are expected to manage higher standards at the same or lower unit cost'.

With 80 Training and Enterprise Councils, a government department may reasonably seek to manage in a meaningful way. So, too, with 100 universities. Probably the same is true with 500 colleges of further and higher education (some LEAs managed more institutions). However, it is inconceivable that meaningful management can come into the DfE's relationship with 25,000 governing bodies. It can only be a bureaucratic, paper-based relationship. That's not good for educational governing, in our view. The gap between the centralized and the devolved is too great.

## Centralized standards, centralized power

The other vein of energy which the government has tapped is the appealing notion of individual standards. This has been manifest in the moves to give each child the entitlement of a common education, in the supremacy given to National Curriculum requirements, in the moves to give to each governing body greater powers of self- and local management whether under the 'government discipline' of grant-maintained status or in the LEAs, and in the moves to give individual parents more and more relevant information and the opportunity to engage individually with their child's schooling. All these seem to us to be moves in welcome directions, quite compatible with the good governing of schools.

The government's twin moves to individualism and centralized power are most obvious in the governing body of its preferred-model school: the grant-maintained or self-governing state school. Parents are the major representative group, at least seven in a governing body of 15 or more; the proportion could be higher. As the equivalent of a business executive director, there is the head, and there are one or two teacher-governors. Nevertheless, the Secretary of State appoints to the category of 'first governors'; these out-number the parent-governors plus head plus teacher-governors. Under the Education Act, 1988, the Secretary of State makes 'such provision as he thinks fit for filling vacancies if it appears to him that the Governing Body is unable or unwilling to fill' any vacancies in first governors. The Secretary of State has 'power to appoint not more than two additional governors if it appears to him that the Governing Body is not adequately carrying out its responsibilities'.

The governing body itself, with its permanent majority of first governors, has the power to appoint subsequent first governors. This, which is quite unlike the LEA-maintained governing body's power to co-opt, seems to us to be a risk to the local representativeness of the governing body, a risk for a self-perpetuating oligarchy. Research, funded by the Leverhulme Foundation, reports that in 1993, 'there are few signs that (GM status) increases the parents' or governors' democratic control of schools'.

## What do you see?

We have offered you some of the things we have seen in the ebb and flow. They are examples of what we have to look at and grapple with if we are to look further, to understand the strengths and weaknesses of the new system of which the new style governing body is a part. Finally in this section about the government's vision and drive, we offer you some questions about what you see:

- Any authority that is not central?
- Any powers that are not delegated and dependent?
- Moves to get back to 'the good old days'?
- What are the pros and cons of expressing everything in contract terms?
- What are the pros and cons of seeing the parent as an individual consumer?
- What is there by way of real consultation, response, second thought and amendment?
- Is there a fair and practicable deal for 1.3 million single parents and 2.1 million children with single parents?

## What are the energies and driving forces at school level?

At school level there are perhaps three major characteristics:

- The winds of change that blow from the government, sometimes with seemingly hurricane force;
- The flow of goodwill and proper interest from the heads, staff, governors and parents for the children of today and tomorrow;
- The cross-currents and awkward situations caused by conflicts of interest and view.

The government has substantially increased the responsibilities, powers and accountability of heads and governors in the governing body. Heads and governors have been and continue to be interested and willing to take up and exercise these responsibilities and powers.

### The flow of goodwill – for the sake of the children

This interest and willingness, for the sake of the children, is the most significant energy and driving force that arises at the local level of the school.

In a parent the commitment to the interests of 'my child' is vital and appropriate, and is echoed in the legal obligation to secure the child's education 'by attendance at school or otherwise' (as the 1944 Act still has it). It is not selfish. For teachers the contracted duty and the professional enthusiasm to create learning and social situations for the children of today is again entirely appropriate, and overriding of all other factors. For the head and the governors in the governing body, however, the children of today are not the only consideration. Tomorrow matters. The interests of the children of tomorrow is a relevant factor in planning and policy-making.

### Driving forces around grant-maintained status

Grant-maintained status may be the issue in the schools sector of education where local energies and forces most dramatically shape the local outcome. Its consideration raises quite acutely the question, 'Is a decision driven by the self-interest of the parents, governors and staff of this school today in fact a selfish decision related to the wider community?' For governors and the head it is possible to weigh the interests of these children of today and of the children of tomorrow. For the parents, however, who make the decision, the responsibility to consider the interests of the children of tomorrow is likely to be less important than the interests of 'my child'. Research into

GMS has revealed the following reasons commonly given for going grant-maintained:

- 'We went GMS for the independence'
- 'We went GMS for the money'
- 'We went GMS because of the way the LEA treated us'.

Yet more than 95 per cent of heads and governors in their governing bodies have not moved to a ballot so far. More than 95 per cent of parent bodies have not asked for a ballot. The government believes that parent bodies may not know about the GM opportunities, that governing bodies and LEAs may be thwarting the earliest enquiries; consequently the 1993 Education Act requires that every governing body must discuss the GM option every year, and must report that discussion to the parent body. Nevertheless, although GM status is a prime policy objective, the government is still leaving it to governing bodies and parents to determine the option and pace.

The parent bodies and governing bodies which have not moved to a ballot on GMS are likely to have had mixed reasons. We suggest some:

- 'We're not up to it, (yet)'
- 'We prefer to draw on one support, which we know, the LEA, rather than have to make a new network of advice and support'
- 'There don't seem to be many providers of advice and support, anyway'
- 'The GM system is still too uncertain'
- 'We don't want to be a school on its own'
- 'We have enough powers, virtually the same powers, in our scheme of local management, so we feel we already are a self-governing state school'
- 'It's unjust and divisive'.

It may be that heads, staff, governors and parents have a loyalty to an educational scene that is larger than a single school. That loyalty may not be to LEAs as they have worked up to now, nor to LEAs as presently constituted, empowered and boundaried. But if that loyalty beyond the particular school does exist, especially if it exists among parents, we suggest that the government will need to respond creatively. GM status may have much to commend it, but so does an education system. We should all be thinking about how to combine the best of both.

## Five cross-currents and awkward features

We referred at the beginning of this section to the hurricane winds and to the flow of goodwill and proper interest. What about the cross-currents and awkward features? We can illustrate what we mean with five examples.

### Overloaded and deferential

The government has given to governing bodies the authority and powers for certain tasks. Governing bodies generally feel weakened and distracted by what they see as an overload of work and responsibility, they have a sense of needing to ask permission. These are not helpful states of mind for self-governing. Later chapters of this book help governing bodies and heads towards gaining a better grip on their roles and their rights. Governing bodies will continue to feel disempowered, even paralysed, while they dance to someone else's tune rather than their own.

### Conflicting interests

The governing body's composition is designed to bring together those with an interest in the school. The interests will quite reasonably be in conflict since the governors come from different perspectives. Instead of having arguments all over the place and instead of having difficulty connecting them to the working of the school, the arguments are helpfully focused into the governing body, as in Figure 2.1.

By definition, therefore, life in the governing body must be expected to be potentially argumentative. It's not meant to be a comfortable ride if the values and priorities are really being explored. Many governors and heads need time to understand this reason for the governing body's existence – to mediate between conflicting interests – and to develop the skills to work through their discomfort.

### Party politics

Even more discomforting is the possible presence of political party politics in the governing body. The Conservative government reasonably sees the governing body as one agent for the implementation of its values and its policies. GM status, a particular pace in introducing assessment and report- ing, and performance-related pay, are three areas (for the sake of example) where other national political views challenge the government's views. If the governors and head choose to see themselves as no more than obedient agents, there may be few significant ripples. But if the governing body

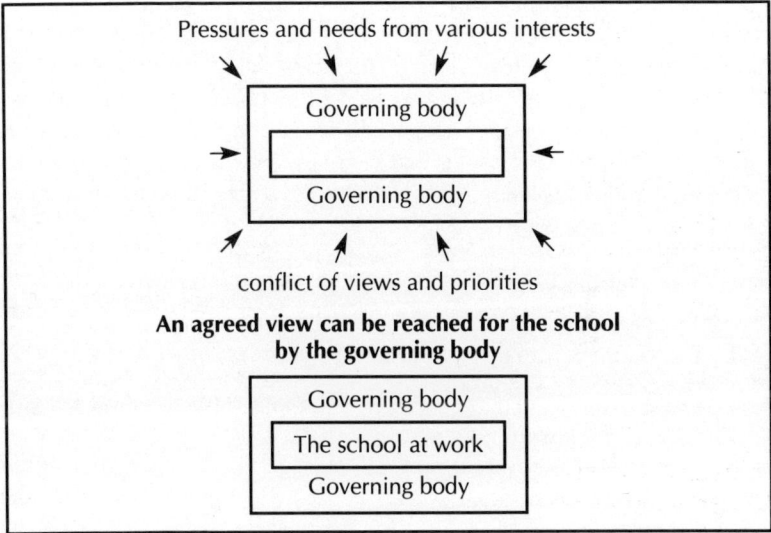

**Figure 2.1** *The governing body as focus for handling conflict*

wants to think for itself, then party political arguments will be a reality in the governing body. And they are the more likely as a consequence of the government's policy of minimal prior consultation. The government allows all of us as heads and governors little time for discussion and general agreement before it implements its policies. In itself the speed may be neither good nor bad; an inevitable consequence, however, is that argument is likely later. Heads and governors are uncomfortable with party political argument. Their reluctance holds back the development of the governing body. It is one of what we call the 'cross-currents'.

*Limited experience of political behaviour*
We began this chapter with comments about a widespread limited understanding of the necessary politics in any organization. Both Conservative Central Office and Labour Party Headquarters already organize briefing and training for governors who give them their allegiance; they see the relevance and the need. All governors hold views on politically-framed policies even if they have no expressed allegiance to a party. Most governing bodies

already have quite skilful political behaviour at work in the relationship of head and chair; sometimes this extends to the chairs of major committees. But most governing bodies have not yet developed their political behaviour in many other areas, such as vis-à-vis the parent body, the different groups of staff, the various employers' organizations and local councils, nor among the governors themselves outside the meeting. The inexperience of governors and heads in this area, and their hesitation as to whether political behaviour is proper, is a hindrance to better governing. It is one of what we call the 'awkward features'.

## The wariness of heads

Our final example of a cross-current and awkward feature is the wariness of heads. The wariness is to do with these key questions:

- Is it really sensible to share authority with governors, to make the governing body the directing forum of the school?
- Is it proper for the governing body to manage the managers, to oversee, supervise and manage me, the head, in any meaningful sense?

Mistakes may be costly to the education of the children; there's a lot of power at stake. The *Times Educational Supplement* columnist and governor Joan Sallis was asked what was her one overriding impression of the past year, with regard to the changes in school management. She had no hesitation in saying it was the striking difference between what heads said about their willingness to share power with governors before the introduction of LMS, and what they did once they had LMS.

Some heads fear that governors may not simply 'join the team', as the heads put it. (It usually means 'my team'!) They fear that governors may take the power. Governors can, may – and wrongly – do get active in matters that are properly part of the head's day-to-day management of the school. This provokes David Hart, general secretary of the National Association of Head Teachers, to argue that governors have far too much power, and to call on the government to curtail it, for example, by giving heads more power in staff appointments and in staff discipline. Hart argues that the accountability of heads personally is much more acute than that of governors. Certainly, in the introductory debate to the 1993 Education Bill, the Junior Minister for Schools spoke of schools that might be 'at risk', and said, 'the sort of firm and effective action which tends to be most effective is changing the headteacher'. And heads may question the appropriateness of their accountability to what some see as a bunch of well-meaning amateurs. At a recent conference one head freely commented, 'we don't want to be accountable. Most of us con our governing bodies'. Governors have to get beyond defining 'volunteer' as 'amateur' in the way they act as agents of accountability, and heads have to recognize that they cannot behave like loose cannons.

Some heads have the same difficulty as do other governors in making sense of the corporate nature of the governing body. For heads the difficulty is made more complex by (1) having simultaneously the role of head as chief executive, responding to discussion in the governing body, and the role of head as governor; and (2) the head's daily experience of parents as *individual* parents, even when assembled en masse for a class occasion, year group or whole-school occasion. The corporate nature of the parent body is a difficult additional concept to develop.

So the wariness of the heads is understandable, is to be respected, acknowledged and worked with. However, left to itself, it acts as a brake on the growth of the governing body and better governing.

## What are the driving forces elsewhere in the system?

When the driving forces from the government are experienced with the force of a hurricane – 'deluged by paper' is how one head put it – it is difficult to detect other potential forces at work:

- parents are not yet mobilized as a force;
- the professionalism of teachers is swamped;
- quangos such as the erstwhile National Curriculum Council have tried too hard to please ministers and experts;
- it will take time for the independent voice of OFSTED to become anything like a driving force.

### Parents

Parents are active as individuals. Some act together at school level, via some form of parent association, usually to raise money. As a national force they have had moments of unity in voicing a request, or making a protest, for example about the format and timing of assessment and testing. As yet they have not been able or inclined to get most governing bodies to listen to them in Annual Parents' Meetings. They have not been stirred significantly by the government's advertising to ask for ballots about getting GM status for their school; most ballots have been called by governing bodies, urged on by heads aware of the financial benefits to the school. The potential for parent-power is still waiting to be realized.

## The professionalism of teachers

The professionalism of teachers has been struggling to come to terms with the requirements of the National Curriculum and its related assessment. The teachers' associations, barred from negotiation for their members' pay, have secured only an appraisal system that allays the teachers' fears of being threatened by appraisal. Pressed by goverment and managed locally by governing bodies and heads, the teachers, far from being revolutionary, largely do what they are told. Such acquiescence, however, tends to produce low morale and a sense of helplessness which depresses real education rather than creates it. 'My interim report is designed', said Sir Ron Dearing, in his 1993 review of curriculum and assessment, 'to increase the effectiveness of teachers and schools by cutting down on central prescription and thus giving teachers the scope to draw on their own professionalism'. To that we can only say hooray and amen.

The recent precedent of teachers, governors and parents saying no to particular forms of testing is an important and welcome flexing of the muscle in the system. It was perfectly acceptable for the government to initiate new plans for providing information to parents, employers and themselves about the progress of pupils; it is equally acceptable for teachers, parents and governing bodies to make it known when they experience the results of those plans as counterproductive. The principle may be right, the mechanism and set of results wrong. Open-minded negotiation should produce a better set of results.

## The National Curriculum Council

The National Curriculum Council (NCC) and its committees have experienced the government's consistent changing of their membership, with moves to the political right and with moves away from those in 'the educational establishment' – that group with which John Major was so looking forward to 'another colossal row'.

Duncan Graham, the Chairman and Chief Executive of the NCC, understood when it began that, 'it was wanted by ministers to give totally independent professional advice, over and beyond the kind they would get from the civil servants'. What he experienced was 'a standard textbook way of sorting out quangos, which was applied to the NCC from the start: strangle the budget, amend or delay the corporate plan, interfere at meetings'. What he experienced was a change of ministers, from Baker to MacGregor to Clarke. He found that the NCC was reduced to being a dependent agency, not an independent force. His view was that, 'unless ideas – and power – are shared, the country will not reap the benefit of the changes'. He warned that, 'a proper evaluation of standards is going to become increasingly difficult as there is little independent statistical evidence available since the politicisation of the NCC and the Schools Examinations and Assessment Council, and the emasculation of HMI'. Such is the perspective of one key insider. The opportunity for something different was there; the Dearing Review and the Government's response to it show that the opportunity may still be there in the new Schools Curriculum and Assessment Authority (successor to the NCC and SEAC). The opportunity depends on the government's capacity to create and let go.

## The Office for Standards in Education

The Office for Standards in Education was set up following the Education (Schools) Act 1992. The Chief Inspector, in addition to informing and advising the Secretary of State when asked, 'may at any time give advice to the Secretary of State on any matter connected with schools', 'may make such reports as he considers appropriate,' and, 'may arrange for any report made by him to be published in such manner as he considers appropriate'.

The opportunity for a disinterested view and voice of education is there. For the moment, the Chief Inspector may be preoccupied with getting each of the 25,000 schools inspected once in every four years. As heads and governors, we should welcome this serious attempt to provide us, among others, with a source of alternative intelligence about our schools.

In conclusion, then, most potential driving-forces seem at present to be acting as passively-obedient agents, or seem to be not yet fully formed. A key question is, 'Are they submerged and holding their breath or are they already drowned and dead?'

We believe most governors, parents, heads and staff are holding their breath while the hurricane winds subside or blow themselves out. Then we can together discern what are the enduring changes for good, and what needs to be cleared away.

## The authors' particular perspective on the changes

It's easy to be pessimistic with a short-term view. The White Paper *Choice and Diversity* was right to pick up the longer-term concept of 'an evolutionary framework'. We accept that systems and relationships, like people, take time to grow. We accept that conflict and contest are healthy signs of life and growth; indeed we wish there were more. With better guidance and more experience all of us as governors and heads will become better practised at handling conflict, be more sure of fair fighting on the issues, be less likely to slip into personal attacks and offensive comments. Taking time to map the conflict, wherever it occurs, helps.

### Mapping the Conflict

Table 2.4 offers a format for mapping. What's needed is to concentrate on the central issue, to identify the different people or groups involved, and to find out and note down their particular needs and fears. And to share the map.

**Table 2.4** *Mapping the conflict*

| Participant A<br>Needs<br>Fears | Participant C<br>Needs<br>Fears |
|---|---|

|  |
|---|
| The issue |

| Participant B<br>Needs<br>Fears | Participant D<br>Needs<br>Fears |
|---|---|

We use three yardsticks to assess whether a proposal is worthwhile in governing:

- Does it make for a greater whole?
- Does it provide and widen responsibility?
- Does it call for and enhance practicable commitment?

So, with regard to the government's general educational programme . . .

## Does it make for a greater whole?

YES in the notion of entitlement, via the National Curriculum
NO in that the weak could lose most – some consumers have less purchasing power
YES in the hope for higher standards
NO in that priorities and funds discriminate – for example, in favour of GM schools
YES in recognizing differences and wanting to offer choice and diversity
NO in not accepting and working with real differences and needs – for example, a common funding formula neglects inner cities
NO in using the economic models of survival in the market place.

## Does it provide and widen responsibility?

YES in measures of local self-management
NO in the National Curriculum's emphasis on being taught and having it inculcated, rather than on learning and developing
YES in promoting diverse kinds of organization
NO in the clumsy mechanisms of statute, regulations and mass-testing
YES in providing for regular inspections and action plans
NO in the centralization of authority and the removal of intermediate structures such as LEAs
YES in emphasizing accountability
NO in under-funding
NO in not developing effective accountability mechanisms – for example, not promoting the Annual Parents' Meeting as a forum for negotiation
NO in government's increasing power to nominate people rather than have them elected.

**Does it call for and enhance practicable commitment?**

NO in restricting parents to the model of consumers
YES in offering explicit values
NO in multiplying the need for everyone to do everything
YES in making real demands on governing bodies
NO in under-funding
YES in encouraging parents to take an interest in and responsibility for their own children
NO in telling rather than negotiating.

What we see, on balance, makes us less than confident for the future, because of the government's interpretations of partnership. The arrangements *could* be made to work beneficially. But the climate for an evolution that is healthy (in our terms) is not evident when a government minister says: 'The Governing Body's role is in delivering to children whatever the system requires at the time'. Delivering is an activity low in authority and dependent in nature. There is potential power in those who deliver saying, 'No, we don't want to deliver (x) or (y)'. We have seen it already in relation to testing and in the determining of a slower pace for GM status. That 'No' *could* bring the government to a different interpretation of 'partnership'.
We like the definition of partnership which we gave on page 20:

> A working relationship that is characterised by a shared sense of purpose, mutual respect, and the willingness to negotiate. This implies a sharing of information, responsibility, skills, decision-making and accountability.

For the partnership to be real, teachers, parents and governing bodies need to formulate initiatives to which the government can say 'Yes'.

# References

40. 'Life is a rose garden', author unknown
41. *The Missing Links*, National Consumer Council, 1986. (See also *Home and School: Building a Better Partnership*, NCC, 1994)
41. Sylvia West, *Educational Values for School Leadership*, Kogan Page, 1993
44. Dr Mike Kelly, at a governors' conference, 1992
45. Research into GMS, Fitz, Power and Halpin *Opting for Grant Maintained Status*, Policy Studies Institute, 1993
47. Research into GMS, *Managing Autonomous Schools*, edited by Bush, Coleman and Glover, Chapman 1993 (and reference 45, above)
53. Duncan Graham and David Tytler, *A Lesson for Us All: The Making of the National Curriculum*, Routledge, 1993

# Chapter 3

# Metaphors and Meanings

---

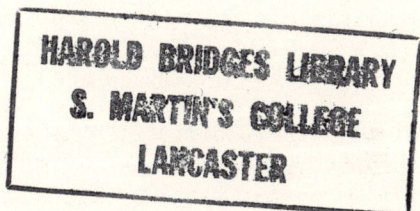

*In this chapter ...*
Introduction – why use imagery at all?
What's wrong with being 'a business in the market place'?
How do others see our contribution?
What pictures lie behind the words the government uses?
How do parents, governors and heads see things?
What images are there of the governing body?
What happens if we think for ourselves?

---

## Introduction – why use imagery at all?

One of the most powerful training activities that we have used with heads and governors is to ask them to draw pictures. Sometimes we ask for a map or diagram, at other times we ask for a metaphor or image to represent how they see things, how they want things to be – their vision.

Imagery, using words in contexts where their original literal meaning does not apply, is part of everyday speech and thinking. Imagery conveys the essence of something; it opens up the possibilities, where literal words keep ideas narrowed down. Imagery helps by loosening up the imagination; it relates to our emotions and to our values. In this chapter we give no more than samples of the imagery around the task of governing schools. We want to illustrate the power of imagery, rather than provide a definitive survey.

Imagery can help the governing body to think more usefully about the school and about its task. Examining other people's language and imagery may provide greater revelation about their motives and vision. We offer the following poem by Elizabeth Jennings as illustration:

## ANSWERS

I kept my answers small and kept them near;
Big questions bruised my mind but still I let
Small answers be a bulwark to my fear.

The huge abstractions I kept from the light;
Small things I handled and caressed and loved.
I let the stars assume the whole of night.

But big answers clamoured to be moved
Into my life. Their great audacity
Shouted to be acknowledged and believed.

Even when all small answers build up to
Protection of my spirit, still I hear
Big answers striving for their overthrow

And all the great conclusions coming near.

What this poem raises for us is the image of a timid person futilely building petty defences against something that she knows will and should take over her life. Indirectly, images of thunderstorms and tidal waves come, too. As an image of the governing body preoccupied with the minutiae of management, while matters of values and policy, evaluation and equal opportunity, move onto the agenda, we know no better!

It is important to remember and distinguish the actual, the direct labels which define what things are and are not. Here are some vital real elements for governing:

- I am *a governor*. I am a *headteacher*.
- We are *a governing body*.
- We are part of *a school*.
- We and our school are part of *a local community*.
- We have a *local education service* for our local community.
- We have a *national education system*.

These are not 'image' words: they mean what they say. They are not loose approximations as Gertrude Stein was the first to say: 'Rose is a rose is a rose is a rose'.

## A health warning

An image is always loose, it shouldn't be expected to fit exactly, nor swallowed credulously. The illumination and the vision which it gives take on the unreality of mirage when we take the imagery at face value, or too far. We may have to settle for a limited clarity and for some uncertainty. Perhaps the biggest hazard is when we unconsciously accept someone else's imagery and start using it without thinking it through. We need to examine where some of the everyday images come from. Some images are more transferable than others, but how are we to know if we have no experience of the field from which the image is borrowed? How many governors have experience of a board of directors which enables them to assess how well this now commonplace image sits on a governing body? Perhaps every image should carry a health warning with it, 'Taken too far, this image can seriously damage your mind'!

## What's wrong with being 'a business in the market place'?

Dan Cruickshank, chief executive of the National Health Service in Scotland, was previously managing director of the Virgin group of businesses. He describes the contrast between the annual general meeting of the Virgin group and his appearance in front of a parliamentary committee as the difference between 'chalk and cheese. An AGM is easier by a ratio of 100:1, in terms of tension, challenge, obligation to answer questions, ability to obfuscate or say "I'm not going to answer that".'

The Hanson group attempted at its 1993 AGM to give the chairman power to stop shareholders speaking more than once on any subject, to interrupt the speaker if the conduct of the meeting was prejudiced, to reject any call for a poll that the chairman deemed 'irrelevant, vexatious to, or inconsistent with the business of the meeting'. It also tried to prevent director-nominations being made by anyone with less than 10 per cent of the shares. A storm of protest caused the withdrawal of the proposals. There's no reason to suppose that Hanson's wish to control is untypical of the attitudes of the managements of business.

## The temptation?

So why might those in education be tempted to use the image?

- It's a closed system, with clear edges, without the web of relationships that an open system needs. It seems understandable.
- People may think of it as everyday, sound, and therefore as acceptable.
- People may believe that the business of manufacturing industry in the nineteenth century, and the business of financial services in the twentieth century, made Britain 'Great'.
- The business model seems to give the school, the head, and the governing body some kudos. It seems to provide a common language when talking with employers and business partnerships.

### The implications

**Certainly, education needs business-like methods**

**Certainly, education does not have like-business aims**

Let us look at some of the implications in the business model:

- Most businesses follow the Marks and Spencer model: they are top-down, are not participative for staff in responsibility nor ownership. More businesses are more like Marks and Spencer's than are like John Lewis. Most businesses are not like John Lewis: characterized by mutual and democratic accountability.
- Most businesses present a culture and working system of individual units with or without local branches. The reality of cartels and monopolies is ignored when education is invited to think of the business model.
- The shareholders and the Board have no real equivalents in education. The shareholders are properly there for the money. The *stake*holders in a school have a much wider range of interests and motives.
- Business is diverse. Which business model are the advocates of the image pushing for its relevance to schools? It seems not that of manufacturing, nor of speculative building and development, nor of the service industry. It appears to be closest to the retail industry – the Sainsbury, Tesco and Marks and Spencer's models, with the notions of individuals as consumers, customers and products, with schools in the market place being driven by market forces.

- We need to remember that many businesses fail. Is education being offered a potential failure model?

There are some other fundamental mismatches between 'business in the market place' and education:

- There's no commitment to stay in business, *but* parents and children need commitment and continuity.
- Demand does not always generate supply, eg in low-cost housing, *but* for a learner to learn, the demands of the learner must be taken seriously.
- All that consumers do is to consume, until the product is gone, *but* education is about relating and retaining. It is life-long.

## How do others see our contribution?

In this section we refer to five models that people seem to have in mind when they think of governing schools, and we reject them all as the definitive model. Each has some truth and some relevance. A governing body is:

- NOT the rubber stamp
- NOT a delivery system
- NOT the local agent
- NOT the big power base
- NOT the company branch.

### The Rubber Stamp?

The rubber stamp is necessary, like a signature on a cheque, or 'received' on an insurance premium; it has importance. It can be a relevant and necessary stage of action. But it is not a participative stage, nor a constructive stage, and the rubber stamp itself doesn't think. The model is usually mentioned when governors feel that the head and staff are not only organizing the school, but also directing, prioritizing, and evaluating it. The governors feel that the head and staff are using the governing body simply as the authorizing validator for professional aims and plans. Sometimes the government tries to use 25,000 governing bodies the same way: 'Oh, but we consulted them. We sent them our proposals in writing. We invited their response'.

The governing body rightly rubber stamps (authorizes) the recommendations from its headteacher appointment panel which does not have full, delegated powers; it rightly rubber stamps the head's termly report before publishing it to the parents; it rightly rubber stamps the contract with the cleaning company negotiated by the special working group of head, vice-chair, and caretaker (co-opted). The governing body isn't the rubber stamp of anybody else, nor for anybody else. It authorizes its own commissioned work.

## A delivery system?

A delivery system is a part of getting something done. Aims and plans have to be brought down to the level of 'How? When? Where? By whom? With what help?'

Implementing is a necessary stage, like filling shelves in the supermarket, like the articulated lorry delivering six new cars to the local garage and its clients. Every system needs its finishers. But it is not often a very creative stage, it is not self-responsible and it has no say in what's being delivered. The model gets mentioned by staff as often as by governors, with the feeling that 'the government decides; we just do what gets pushed down the line'.

The governing body rightly helps to deliver in ensuring that the school 'promotes the spiritual, moral, cultural, mental and physical development of pupils at the school and of society; and prepares such pupils for the opportunities, responsibilities, and experiences of adult life' (as Section 2 of the 1988 Education Reform Act requires).

The governing body rightly delivers to the parents an opportunity for discussion of the governors' annual report, the discharge by the governing body of their functions in relation to the school, the discharge by the headteacher of his or her functions in relation to the school, and the discharge by the LEA of their functions in relation to the school, (as Section 31 of the Education No. 2 Act, 1986 requires). The governing body rightly delivers the spending of 'any sum made available to them in respect of the school's budget share for any financial year as they think fit for the purposes of the school' (Section 36, Education Reform Act 1988). But those last words – 'as they think fit' – show that this particular action is meant to be participative and self-responsible. That's hardly just delivery.

## The local agent?

The 'local agent' goes with a central directing authority. It is necessary, for keeping in touch with local circumstances, that the local agent has delegated authority to make some decisions, 'as they think fit'. The National

Curriculum Council seemed to regard schools and governing bodies as local agents. Yet the NCC and the governing body together tended not to spend time on the local values, local needs, local mission, local aims and local curriculum. The 20 per cent margin allowed by the Dearing review for schools to broaden their curriculum at their own discretion may go some way to redressing this unbalance if governing bodies overcome their reticence to deal with curriculum matters. The NCC could not, and SCAA cannot, 'police' the system. Only the governing body and the body of parents can monitor developments, progress and omissions for each school on a regular basis with some help from OFSTED every four years.

Generally the central directing authority establishes the local agent, may at its discretion cease to maintain the agent, and may or may not state the grounds for its decision to stop. There is usually no appeal possible to a higher authority to challenge the decision. Regrettably, in our view, this means that the governing body of each school, and GM schools in particular, could be an agent of the Secretary of State in a way that is quite different from the relationship of an LEA-maintained school and its governing body with the LEA. The ultimate dependence of GM school governing bodies on individuals – the head locally and the Secretary of State nationally – is, for us, not the model of a responsible adult, not the way in principle a governing body should live and work. We come back to this issue of GM schools later in this section.

## The big power base?

The big power base model is one that may be operating in front of our eyes so obviously that we don't notice it. The big power base assumes that there are no significant other authorities, that its own will is (by definition) reasonable, that its motives are generous, that negotiation is an unprofitable (therefore unnecessary) use of time because it will bring no wisdom and no need to revise a plan. Is your governing body *seen* this way? *Could* it be seen this way, by some staff, on some issues? By the parents? In our view recent governments have at times actually *behaved* this way. In our view it's not the way for governing schools, not the way for the responsible, adult, governing body.

## The company branch?

The company branch is a variant of the local agent. We have already referred to this model as a mirage. Yet it is more influential and seductive, because the government offers business as a model for some things in the education service. We as heads and governors, most of us not having had personal experience of directing and managing a company, may make the mistake of forcing the lessons from business to apply too widely. The Marks and Spencer's branch, for instance, properly offers centrally-designed and centrally-produced goods, works to a central policy, and is on a central national database for stock and daily sales. It has a local manager to recruit and manage the staff. We don't see the governing body as properly being simply the branch manager on behalf of central government. There are too many local variants.

## What about Grant Maintained schools?

We said earlier that the current dependence of GM schools on the discretion of the Secretary of State was, for us, not the way in principle a governing body should live and work. The duty of the Secretary of State is to maintain (that is, the duty is to make payments), with the amount allocated in accordance with regulations made by the Secretary of State. The governing body is incorporated if the original proposals are approved by the Secretary of State; the Articles of Government are made by order of the Secretary of State. The Secretary of State may cease to maintain by giving notice of his intention and may give the governing body a notice stating his grounds (there is no appeal against the Secretary of State). In principle, this seems absolute dependence. That is not to say that authority isn't delegated to the governing body to act. In most cases, that will mean the delegated powers of local management, plus some additional powers delegated to the governing body of a GM school. Our purpose here is not to have a general argument about the pros and cons of GM status as such, but to underline how the powers of creation, funding, direction and dissolution by the Secretary of State make a GM school governing body more of a dependent agent, less of a local authoritative body. This conflicts with our view of what's best in governing schools. We believe in less dependence, more authority and accountability as local citizens.

The governing body of the grant-maintained school:

- probably has greater influence, at the margins;
- does not have much more control of money, although it may, at the government's discretion, have more to control;
- probably can make an advantage out of being the employer;
- will probably be expected to embody for the school the government's

five great themes from the White Paper *Choice and Diversity*: quality, diversity, increasing parental choice, greater autonomy, and greater accountability.

We have commented above about greater autonomy. We see it as the rhetoric. We don't find it, in practice, much different from LMS, and we do fear the legal dependence. We look at accountability in a later chapter.

## What pictures lie behind the words the government uses?

The use of language and images springs from the imagination, the feelings and the values of those who use them. The words the government uses – either through its ministers and its officials, or through what it publishes – relate to what really matters for the government. We have chosen the 1992 White Paper to illustrate how the Government sees education. We hope that the example will encourage you to look more closely yourselves at speeches and documents emanating from those who shape education.

The imagery in *Choice and Diversity*, the 1992 White Paper which preceded the 1993 Education Act, is often mechanical and technological; sometimes fluid and from the natural world; minimally of growth and human life.

The mechanical is in *the framework,* the *drive* for higher standards, the *parent-driven* nature of selection, with education *geared* and governing body decisions *geared* to particular needs. GM ballots are described as a further powerful *instrument*. School days are foundations for *building on*. There are *central pillars* of educational architecture. There is *impetus, momentum, keys, obstacles erected*, policies set *in train* and attendance as the best *litmus test*. Parent-influence is *entrenched*, *pressures* need to be met with *bulwarks*, and children need to be *stretched*.

From fluids and the natural world, we have aims that *cascade down*, legislation that *flows, channels* for complaint, schools in the *mainstream*, and young people *drifting* into a life of crime. Parents *flock*, talents are *harnessed*, and things are *pinned down*.

What images are there that draw from the reality of human life and growth? There's *ownership* through GM *status*, there's *fostering* through testing, and there are firm *marriages* – the government is *firmly wedded* to quality, and to parental choice and involvement. Power is in *hands*, schools turn a *blind eye*, (but there are new and powerful *watchful eyes*). There are *first steps* and there's the *heart* – *the heart* of government policy, and proposals at the *heart* of legislation.

We have three agricultural images depicting growth and change in *finding a root*, *transforming the educational landscape*, and *strength in a particular field*. Yet surely the models that most suit education are of growth, of organisms, of development. What matters to the government, on this sample of its language, is a more mechanistic set of values.

## How do parents, governors and heads see things?

In 1992 the DfE funded some research into why Annual Parents' Meetings were not working. Part of the work involved interviews with heads and chairs, and with parents, some who came to APMs, some who did not. We offer the imagery which occurred spontaneously in those interviews as an indication – and probably as a fair illustrative sample – of their current feelings about life in the world of education.

That life seems constricted. Governors are 'bogged down in interpreting legislation'. 'My ability to negotiate is being stifled'. 'It's a trap, really'. 'The chair wants everything very cut and dried, organized and controlled. She's nervous of questions from the floor'. 'A "dry" report won't break down the barrier'.

The values of public life seem few and narrow in scope. 'Schools are bound by tradition'. 'The chair is carrying ten years' baggage with him'. 'My contribution as head is to keep things ethical and above board'. However, 'We have parents now fighting to get on the governing body'. 'Let's get the issues aired publicly'. 'Education and this democracy in school go absolutely hand-in-glove'.

Power and authority seem to be experienced as measures of aggression and control, rather than as resources for enabling and educating. Heads will have their 'knuckles rapped'. Governors 'in the APM are up against the firing wall'. Both are 'wary of the way things can get steam-rollered out by the minority'. 'Parents don't need a formal invitation in which to throw bricks at us'. 'The chair is a law unto himself'. 'You don't want "professional" pushed down your throat'. 'I don't want the teaching staff on the block, so I don't encourage them to attend'.

The 'system' that includes the APM, the parents, the governing body and the head, seems mechanical. 'Feedback', 'channels', 'tools', 'breakdown', 'vehicle', 'steady state', 'leverage', 'taking off', 'taking the temperature', 'sounding-board', 'the interface', 'the yardstick', 'the model', 'the nuts and bolts of the school's operation', 'the head of steam', 'the hot-bed' – these are the single largest group of images, about a fifth of the total. The 'system' is less often seen as some kind of organism, capable of growing. There can be 'festering'. 'Our reports go out ... "warts and all"'. 'A new breed of governors is needed'. 'This governing body has a distinct personality'. 'The grass-roots have to influence policy'. So strongly has the business imagery been pushed that even here the relationships of parents, governors, staff and others are seen as predominantly those of commerce, partly like those in other areas of public life, barely at all in personal terms. The imagery that predominates is of 'shareholders', 'taking stock', 'takeover', 'ownership', 'consumers', 'shop-keepers and customers', 'on display' – that dominant business imagery again! Just occasionally there was personal imagery: 'School is a bit like a family', 'the blind leading the blind', 'the shepherd'.

Finally, we offer two key images which bear fundamentally on government and on governing bodies: 'It takes a long time in education for something to become part of the fabric,' 'It's the DfE – rather than the LEA – that now has the stranglehold over schools.'

## What images are there of the governing body?

Look at what the group of governors and head is called – the governing *body*. It took us, as authors, some time not to ignore the obvious, to realize the potential of the imagery in that word 'body'. Each body is unique, has its own life, the different parts need to work together, it thinks, it acts – whatever other images, helpful and unhelpful, are highlighted in this chapter, we suggest you hang on to the image of the body.

### The low-key images of the governing body

The most common images are:

- encouragers
- supporters
- companions
- hand-holders
- and (with some ambivalence) guardians.

Many heads and governors prefer the low-key images and style. A survey in 1993 of nearly 500 governors and heads in Warwickshire found that 'it is felt that by far the most important function of the Governing Body is "to support the work of the school" while its role as "critical friend" or "mediator" is not experienced to be very significant'.

When training with groups of heads and governors, we ask them to come up with as many role words as possible. They usually produce a mixture of the more traditional low-key activities of the 1960s and '70s, and others that reflect the increased responsibilities of the '80s and '90s. The low-key, low-esteem imagery, often said in a negative tone, include 'the government's delivery boy, cypher, scapegoat, rubber stamp, lamp-post, Victorian wife, juggernaut, a custard pie thrown at the school, jury, politicians, an amateur society, a collective'. Kenneth Baker, as Secretary of State, spoke of 'the sledge-hammer'.

## The more assertive and positive images of the governing body

The most common images are:

- advisers
- initiators
- enablers
- and guardians.

Other images that we have come across include – 'a Cabinet (with corporate collective responsibility), bridge, champion, strategic planner, partner, contracted agent, think-tank, ombudsman, representative, cultivator, key, generator, musical band, foster-parents, a walking party, a tunnel with windows, a board of trustees, a corporation, a womb and a body, a beehive, a cooperative, public relations officer, personnel officer, a manager, an accountant, a judge, an advocate, an overseer, a watchdog, an arbiter, a learner, and (above all), a *body*'

We suggest that one of your tasks in your governing body is....

- to find the images that work for you

We suggest that you think about:

- What does giving service mean for us as a governing body?
- What kind of a system do we think we're in?
- What do we want to do with our body?

We suggest that you think about, and reject, the image of the board of directors.

# What happens if we think for ourselves?

Do you know the game 'O'Grady says', also known as 'Simon says'? It's a game traditional among children, and used in army training for its power of sharpening the wits. Players obey commands only when they are prefaced by the words 'O'Grady (or Simon) says'. Other orders must be ignored. Speed of utterance by the leader is, of course, essential!

## THE VOICE OF AUTHORITY

### A Language Game

Do this, Don't move. O'Grady says do this.
You, get a move on, see, do what I say.
Look lively when I say O'Grady says

Say this, Shut up. O'Grady says say this.
You talk fast without thinking what to say.
What goes is what I say O'Grady says ...

and the poem goes on to Kingsley Amis's neat conclusion:

'... O'Grady must be out of this
Before what we say goes'.

Others can offer; can require and can instruct. They can't make governing bodies play the game. Governors can walk away at any time; that's not easy for heads. Governors are volunteers, but that does not mean to deny that what governors in the governing body have is responsibility and accountability. All their discussions and decisions and actions add up to an overall impression, a picture, of the job as they are making it.

Try drawing pictures of your governing body's work and its connections. Get your images into your words. One governor we know drew a beehive with all the bees busily engaged about their business. Most were happily contained within the hive but one or two were doing their own thing at a distance. An interesting discussion ensued as we drew out from that governor whether those working at a distance were on commissioned business or acting as mavericks! A map is another helpful representation of where things are in relation to each other. It gives illumination and insight. But maps can be deceptive and therefore hazardous when we don't know the purposes for which the map was drawn and when we don't know the principle on which some features were selected and represented and some not.

Try mapping the different groups, pressures and aims. The illumination and vision get confused when we forget that the map does not tell us about:

| | |
|---|---|
| WHY to travel | WHERE to travel |
| HOW to travel | WHEN to travel. |

What would be best is a model and description that represents a number of mutually supportive and not subordinate posts or groups, **in balance**.

Think again what you mean by the school (look back at Figure 1.3 on page 27).

Don't narrow your meaning to include too few people. Don't say 'staff' when you mean just 'teachers'. Don't narrow it to mean just 'academic learning'. Remember the child's *life* and learning. Remember that we learn from an encounter with someone, not just by being taught or told.

Use your values as criteria. We suggest the following to help you:

- Does this promote a more intense awareness of something greater than we've already got?
- Does this get people to make personal and responsible choices?
- Does this bring out in people a sense of commitment and giving, rather than an over-burdened 'sense of duty'?

So, what happens if we take this thinking into our work? Some imagery can help governors get more understanding of their purposes, tasks, functions (getting bearings, taking soundings, setting a course).

Some imagery can help governors get more understanding about their behaviour, processes, working methods (helping things grow, having an open door (and ear), the balances of power).

## The scale of it all

Bodies have functions and governing bodies have roles and functions based on a wide range of responsibilities. Chris Lowe in *The School Governor's Legal Guide* reckons that there are over 800 different legal responsibilities. We are going to need a yardstick by which to determine what to do and when in *this* school. The image that we have of ourselves and for ourselves is a helpful starting point. Are we functioning as a healthy, independent-minded, quick-thinking body or are we more like puppets, manipulated by someone else's thinking, or even dummies, virtually non-functioning?

The avalanche of paperwork and directives can easily result in paralysis of the body rather than empowerment. For a body to gain energy it needs to absorb pressure, not simply be overwhelmed by it. Each governing body needs to set its sights on the educational experience of the children. Their life and learning are affected for better or for worse by the imagery around the school, by the models they see.

At a recent conference of primary headteachers, one head protested at the suggestion that there were occasions upon which parents should be encountered as a body. 'They only lead each other on', she said. We suggested that she was in fact borrowing and using the model so frequently used by those in authority, of dividing and ruling, yet in the same school the language of cooperation was frequently used as a model of good behaviour for the children. The head did not like having the inconsistencies pointed out to her. If we behave according to the government's model of giving power to the corporate body of governors we must find ways of exercising it together and encouraging others to do so.

The chair of governors of a Nottinghamshire comprehensive explained where such cooperative working begins for them: 'We're never afraid to be self-critical in our school' – *not afraid*; *self*-critical; *in* our school; *our* school. They use the cycle of development planning as an important mechanism for involving the governing body and the wider community. The process involves:

- auditing, taking stock with the governing body; acting as *eyes* and *ears*, bringing other information and perspectives;
- planning dreams and prioritizing practicalities, with the governing body as *co-planners*; part of the *brain* work;
- evaluating how it is going and has gone, with the governing body as *guardians* and *trustees*.

The governing body has to make sure not only that it functions healthily as a body but that it both holds others accountable and is itself held accountable for the healthy functioning of the whole school. Why and how it does this is explored in the next few chapters.

## References

58. 'Answers', Elizabeth Jennings, from *Collected Poems*, Macmillan
59. Gertrude Stein, *Sacred Emily*
59. Dan Cruickshank, quoted in *Times Educational Supplement (TES)*, March 1993
66. DfE research project on Annual Parents' Meetings, unpublished working papers of Hinds Education and Review and Birmingham University School of Education
67. Survey in Warwickshire, Margaret Maden, *TES*, June 1993
69. Kingsley Amis, 'The Voice of Authority', *Critical Quarterly* Supplement, 1960

# Chapter 4

# Wider Public Ownership

---

*In this chapter ...*

What does 'ownership' mean?
Whose is the school?
– what does the government say?
– how is it really?
– how might it be?
Whose is the governing body?
Taking and making ownership.

---

## What does 'ownership' mean?

Most of the recent change in education is predicated on the rights of parents and the dissatisfaction of at least some of them with limitations put upon their contribution by the professionals. Now there are those who feel that the pendulum is swinging too far in the direction of parents whose behaviour may demonstrate apathy – 'Do the parents really want this?' asked one newspaper columnist in 1992 – or in the direction of those whose own needs may obscure the needs of their children. Could the right of a parent to withdraw their 15-year-old from sex education according to the 1993 Education Act bring them into conflict with the needs of the child?

We believe this passage from *The Prophet* offers an important perspective:

> Your children are not your children.
> They are the sons and daughters of Life's longing for itself.
> They come through you but not from you,
> And though they are with you yet they belong not to you.

> You may give them your love but not your thoughts,
> For they have their own thoughts.
> You may house their bodies but not their souls,
> For their souls dwell in the house of tomorrow, which you cannot
>     visit, not even in your dreams.
> You may strive to be like them, but seek not to make them like you.
> For life goes not backward nor tarries with yesterday.
> You are the bows from which your children as living arrows are sent
>     forth.
> The archer sees the mark upon the path of the infinite, and He bends
>     you with His might that His arrows may go swift and far.
> Let your bending in the archer's hand be for gladness;
> For even as He loves the arrow that flies, so He loves also the bow
>     that is stable.

Parents hold their children in trust. Trusteeship suggests a responsibility for the future as well as for the present. Parent-governors, for instance, can continue after their child has left the school. Trusteeship is one key element for all those involved in sharing *this* particular kind of ownership.

Participation is another element. In *Moving to Management*, Angela Thody demonstrates the recent widening of the ownership of the education service, and illustrates the potential for wider active participation. We believe that the more active participation of some 300,000 governors is already serving to rectify some of the distortions of view previously built on ignorance.

Identification is a third element of ownership. 'We're never afraid to be self-critical in our school,' said that Nottinghamshire chair of governors. The sense of identification conveyed by the words 'in' and 'our' earns governors the right to be constructively critical. That identification comes partly from contributing to the school in action, and from the governors having a sense of belonging to the school.

Ownership has several other dimensions, too:

● the pride of seeing my child in the school play

- the rush to protect my child from the bully
- the stakeholder whose interest is not necessarily financial but is an investment
- the shareholder risking capital and seeking some gain
- the representative concerned for the interests of many
- the responsibility of the user to see that this part of education remains serviceable and not faulty or even dangerous
- the employer giving consent to the employees.

We, as heads and governors need to think, 'Whose is the school?', and 'Whose is the governing body?', and what the implications are of this particular kind of ownership.


## Whose is the school?

In one sense, the school is

> **The pupil body**

Parents, teachers, governing body, local community and government would all acknowledge that essential dimension. What they might not all acknowledge is the sense that the school is

> **The parent body**
> > **the pupil body**
> **as co-educators**

However, they would certainly agree in acknowledging the dimension of the staff as a specialist part of the school

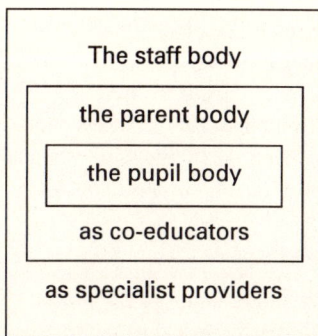

A final dimension is the governing body. Many parents may not recognize the governing body as part of the school; some teachers see it as marginal; some pupils don't know that it exists! Nevertheless, we see the governing body as intrinsically part of the school as Figure 4.1 shows, while having its distinctive purpose and nature

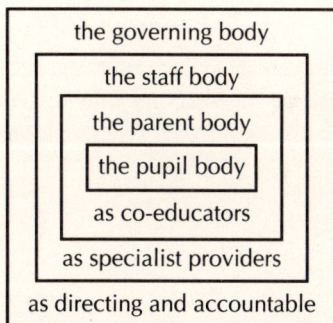

**Figure 4.1** *The school of four bodies*

*Your* definition of the school will affect how you define ownership, responsibility and accountability for the school. It's worth getting all these definitions agreed by the governing body – and by the staff, and by the parents? That would help the parents to hold the governing body accountable from *within* the school, in a context of mutual responsibility, with the government – and, in most cases, the LEA – holding the governing body accountable from *outside*.

This defining of the school matters for the governing body, because such definition will affect their direction and priorities. The definition is as fundamental as the foundations of the school building. The owners of the different kinds of school – the voluntary aided, government grant-maintained (GM), independent, city technology, LEA-maintained – each has a direct stake and different values in setting the direction and priorities of the school.

## The government's view

The government has been keen to stress the autonomous nature of each school within the limits centrally defined. The government has arranged for local interest groups to be represented on the governing body so that it can provide a focus in which to reconcile their different claims. The pupils are not represented because from the government's point of view the school is *for* them; it's not theirs. Nor does the government rate highly the demands that schools might make on each other, their collective responsibility and mutual ownership of education. But the collective dimension is real, with its advantages and disadvantages, as this example from the FE world shows.

### Colleges slip into debt

Almost one in three further education colleges will be in the red by the end of their first year of financial independence, says the Further Education Funding Council, after a detailed analysis of the strategic plans of all the colleges up to 1994.

A total deficit of £42 million will be shared among 102 colleges.

The government has significantly narrowed the meaning and authority of local ownership. It has done this through its insistence on delegated LMS, and through promoting its offer of GM-status. The local community's stake in schools is essentially narrowed to the parents on the governing body and to co-opted persons, particularly, says the government, persons from the local business community. As the government sees it, there is no need for the school to belong to the LEA. The LEA has duties to the school, but no ownership. The LEA, under the 1993 Act, is no longer required to have an education committee. The government does not value local politicians. 'This is a great measure of de-politicisation in education at the local level', said the Secretary of State in launching the White Paper *Choice and Diversity*. The government does not link the school with the citizen, the vote, the citizen's responsibility, nor the citizen's potential interest in the school in the neighbourhood.

## State schools

The government's favoured phrase is 'self-governing state schools'. The government controls the budget total and announces it a year at a time. Sir Peter Newsam, Director of the Institute of Education, wrote:

> If anyone doubts the degree of absolutism that is being put into statutory form, the question to be asked is this: 'What is it in relation to any GM school that, under the legislation, the Secretary of State will not directly or indirectly do?' Open it, close it, enlarge it, change its governors, move it, inspect it, tell it what to teach – all those things are to be in one person's hands ... It has been for an historian, Lord Skidelsky, to ask the essential question: 'How is it possible to reconcile the Government's desire to make parental choice paramount with their vesting of virtually all decisions as to which schools are to open or close, expand or con-tract, in bureaucrats and in politicians?' (Hansard, 23 March 1993). The answer is that it is not. The only thing to do with absolutism is to recog-nise it when it appears.

It seems to us that the reality the government wants is rather different from what the government says. Our impressions of what the government seems to want are:

– we, the government, have the say,
– the head and staff operate the school,
– the local agent of accountability, the governing body, monitors the arrangements and satisfies the parents.

## The government and parents

For the government, parents matter. Balloting for GM status, more power to choose, more power to complain – these are elements of the new deal for individual parents. The government seems to be in touch with something that is important for parents. Victoria Neumark, a columnist in the *Times Educational Supplement*, catches something of this:

> My child needs to be safe enough to be able to work without disruptive kids in the class, with sufficient company of her academically-equal peers, with enough stimulus from her teachers to develop her intellect, and with enough achieving role-models to show her that work is worth-while.

The government has given the Annual Parents' Meeting to the parent body as a potential vehicle for partnership and for accountability. The government has then (inadvertently?) diverted attention from the APM as a forum for discussion of values and possibilities, by offering the governors too narrow a prescription for the governors' annual report, which then comes to dominate and stifle the APM. Heads and governors have not been ready to listen, yet, to the parent body. When the time comes that they are ready to listen, and when they engage in planning with the parent body, the government may find that the APM acquires teeth and bites the hand that set it up! The parent body may lobby the local MP on the level of resources for schools; the APM and the governing body may jointly lobby the DfE.

The parent body's giving of evidence to the registered inspector before the four-yearly inspection, the parent body's receipt of the summary inspection report and of the governing body's action plan, are further opportunities which give it a significant role in ownership, *part*-ownership, alongside the governing body, within the school. The age-weighted-pupil-unit, as the key element in the funding formula for schools, can be thought of as a piece of capital which the parents invest in the school, by their choice of it, which then earns or attracts revenue funding to the school.

## Other views of ownership

Each governing body has a stake in the *whole* education system. What governing bodies collectively have yet to do is to find some way to articulate their views and their energies and to establish their representatives in dealing with, and negotiating with, government, the other major stakeholder, the source of much authority and the wielder of most power.

In practice, at present, most governing bodies spend all their energies in and for the individual school and its present needs. Governing bodies seem not to grasp, make, and use the opportunity of a development plan as one focus for exercising their trusteeship for the children of the future. The head-inspired or governor-inspired move to a ballot for GM status is one of the few forward-looking activities of governing bodies. They seem otherwise preoccupied with one-year budgeting and planning, and trapped by a flood-tide of immediate work. They administer, react and operate, rather than own and plan.

In practice, at present, most governing bodies seem not to see themselves as representatives or trustees for the local community. The government's focus on the parent body, not the governing body, as the determinant for GM status reinforces that limited view, so that governing bodies are a long way from seeing themselves collectively as representatives of the nation's concern for education.

In practice, at present, most heads refer to 'my school', and many may mean 'mine only'. The time they put in, the involvement of their self-esteem with the achievement and success of the school, the effective daily operating which they have to manage, make the 'mine' fully reasonable, understandable and necessary. The 'mine only', the 'why should I report to this lot, the parents, the governors?', is not reasonable, nor is it an attitude of accountability that befits a professional.

In practice, at present, most PTAs, parents' associations, home-school associations and associations of friends of the school, refer to the school as 'our school'. That's probably the 'our' of commitment and of pride, of stake-holders. It probably is not the 'our school' of responsible part-ownership. The wider local community may share the pride, but probably does not share the commitment, even where the school is a symbol of local community life.

## The potential for governing body ownership

We need to see what's there in potential, as well as what's there in practice. The potential is there for the governing body to give the lead to head, staff, parents and pupils, to make 'our school'. The opportunities and mecha-nisms exist for authority to be taken, for responsibility to be accepted, for accountability to be exercised fully and openly. The potential is there for a national body that expresses for all governing bodies what they collectively share. That body would be central to the processes of balanced negotiation between those who have the power to appoint and to get things done (or left undone) locally – the governing bodies; and those who have the power to set national tasks and to allocate national funds – the government. Figure 4.2 indicates the basic relationship. In practice, at present, the body that directly represents the 25,000, and links on their behalf to government, is missing. The present governors' organizations – the National Association of Governors and Managers (NAGM), Action for Governors' Information and Training (AGIT), and the Institute for School and College Governors (ISCG) – do not have that direct representation. And does it all matter, for the children? Are we getting carried away in a new political game?

## Ownership and the children

We think that ownership is vital for children's learning and growing. We think that the wider, richer, deeper, more democratic participation is in the

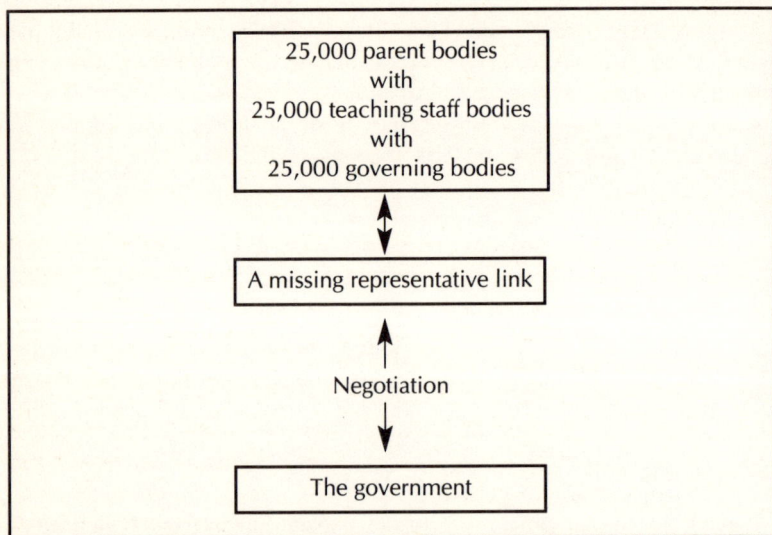

**Figure 4.2** *A missing representative link*

governing of schools, the more likely the schools' working principles are to include the vital participation of the pupil as partner in planning, managing and benefiting from the curriculum.

The criteria we offered earlier, by which to assess the good governing of schools, were:

- Does it advance a greater whole?
- Does it increase and widen responsibility?
- Does it call for practicable commitment?

These criteria coincide with our personal experience of 'good' learning. As former teachers we reckon that if it's good for children, it's good for governing and vice versa.

The Taylor Committee of Enquiry was set up by the Secretary of State in 1975 to review the arrangements for the management and government of maintained primary and secondary schools in England and Wales. It reported in 1977 in *A New Partnership for Our Schools*:

We consider it necessary to propose that all the parties should share in making decisions on the organisation and running of the school since, in our view, this is the best way of ensuring that every aspect of the life and work of the school comes within the purview of the interests acting together.

It's that unity in the life and work of the school that leads us to see as inter-connected the children's ownership of their education and the adult's ownership of the school.

But on the debit side, at present, the government believes that the 'vital educational imperative' of the National Curriculum can be met by legislation. There is little or no vitality welling-up from the hard-pressed and much-moulded education service. There are no signs yet that the present forms of local management and GM status have increased the learning achievement of children. 'It isn't just Brussels that rolls out the red tape. It's Whitehall. And Town Hall. Everyone likes to tie another knot' (as John Major put it at the 1992 Conservative Party Conference). Knots don't help unravel the mysteries of learning.

There is potential, on the positive side, to get the common principles of good learning and good governing moving forward together. Governing bodies have not been convinced by the government's legislation and exhortation about certain ways and times for testing children, and have acted on their own judgements. Governing bodies may learn to harness the power of the government for common, agreed ends. Once heads are comfortable with LMS and GMS they will be in a position to introduce similar, participative, decision-making with their staff (and pupils?). A government might be persuaded to write pupil governors back into the framework. There is evident sense in the principle of participation where one is reasonably affected; parents will find their ways to contribute their say more effectively. The key questions for governing, as for learning, are:

- How long will we all rest content with a role of being dependent?
- How soon will we all use our energy to create an inter-dependent partnership?

## Whose is the governing body?

It's important as a governor or head to find personal ownership in the governing body. It's important for the governing body to be self-conscious, self-possessed, self-critical and to find its self-ownership. The 1993 Act makes the governing body corporate, emphasizing it as an entity rather than an aggregate of individuals.

Research among heads has established that at present the great majority see the governing body as 'mine to manage'. If 'manage' means influence and educate, that's fair enough. If 'mine' means to tame, dominate, control, that's not acceptable. That could be why we have some reported rows between heads and governors – and there may be many more *un*reported. At least one chairman, writing in the *Times Educational Supplement*, 'resigned because I was tired of the perpetual wrestle of power with the Head'.

Accepting mere dictation about tasks, governing bodies and schools could become agencies run for the government and by its officials. The governing body as a mechanism could be swept away in the drafting of a new Education Bill. Following the 1993 Act it already can be replaced with another agency, if it 'fails', ie, by an Education Association. A 'failing' school, a failed and disbanded governing body – the next step is automatic GM status, with no ballot of parents. That scenario, of the governing body as agent, and with the possibility of the scrapheap, is not an encouraging one. If the governing body can:

- find its authority, in representing local parents, the local community and local businesses;
- develop its powers and its partnership with the head, staff and parents, by getting to know, by demonstrating that it cares, by giving time, by determining aims and priorities;
- think wider than the school and see its fellowship with 24,999 other governing bodies and help create a national representative body to negotiate with government;

then it finds its own ownership and accountability. The government's language allows it. How might we as heads and governors go about it?

## Taking and making ownership

What is needed is a new attitude that serves notice to the local and national citizens: 'Under new ownership'. Remember that the government speaks of 'autonomy'. The government is prepared to provide only limited support and is determined not to manage. Regulations and inspections are all that heads and governors can expect. Remember the definition we offered about taking and making ownership; it's repeated here:

- Find your authority.
- Develop your powers and partnerships.
- Think wider than the school.

Essentially our advice is **do it your way.**

We shall spend the rest of the book expanding on that advice. Think of the school with a boundary zone around it. It's a kind of psychological-cum-legal boundary zone. Just the fact that it is a particular school creates that outer edge of identity.

As an individual governor or head, you come from a local community of some kind and in that sense you are representative of that local community. What you bring to the governing body, what you say inside, what you do inside, brings in something of the local community. What you say and do outside carries something of the governing body and the school. The governing body as a whole receives all that its head and governors bring in across the boundary zone. Its messages to the outside world consist of all that the head and governors say and do outside. That representativeness and representing across the boundary zone gives authority to the governing body. And the task for which it has that authority is the better education of the children. All the government does is to be *another* channel of comment about that task; the main channel is the mind of the governing body.

So, in the working of the governing body, in its meetings and those of its authorized groups and committees, in the informal dealings of the head and governors between meetings:

- think and say in the school what you think is important in the local world outside of the school;
- think and say in that local world what you think is important in and for the school;
- use all your skills and courage to insist that the governing body listens to and talks with the outside world in addition to the pupils, the parents and the staff who are inside;
- give 10 per cent of whatever time you are able to give as a governor to contacts, reading, visits and training wider than your school. Remember it's the width and richness of all that you bring across the boundary that matters for the children. We're not suggesting 10 per cent *more* time. Work out what you can give, and give 90 per cent *directly* to the school and the governing body.

If you can find your personal authority, develop your personal powers and partnerships, and think wider than the school, that's the biggest influence you can have on the governing body. Keep your eye on the task: the better education of the children – that justifies all; that gives you your authority.

## Not Individualism

In his 1990 Reith Lectures, Rabbi Jonathan Sacks highlighted the value of wider public ownership over individual consumerism.

> In recent years the key word in our political vocabulary has been the individual. In the 1960s the state retreated from the legislation of morality. In the 1980s it drew back from the economy and from welfare, and it was assumed in both cases that public responsibility would be replaced by private virtue.... We would still be pained by deprivation, but we would address it through self-help and philanthropy. Private virtue was the building that would stay standing once the scaffolding of the state was removed. But without the communities that sustain it, there is no such thing as private virtue. Instead there is individualism: the self as chooser and consumer. And the free market can be a very harsh place for those who make the wrong choices. The shift from state to individual at a time when our communities have eroded has carried a high cost in poverty, homelessness, broken families, and the drugs, vandalism and violence that go with the breakdown of meaning. In an individualistic culture, prizes are not evenly distributed. They go to those with supportive relationships. To those, in particular, with strong families and communities.

In the 1990s, the governing body is a definite focus of the strong public support that a school needs, a demonstration of ownership that goes wider than the self. It can play an important part in making real the possibility that the whole community, weak and strong, can make right choices.

## References

73. *The Prophet*, Kahlil Gibran, 1926, Penguin 1992
73. *Moving to Management*, Angela Thody, Fulton 1992
76. 'Colleges slip into debt', *TES*, July 1993
77. Peter Newsam, *TES*, April 1993
77. Victoria Neumark, *TES*, 1993
84. Jonathan Sacks, *Reith Lectures*, BBC, 1990

# Chapter 5

# Where do the Bucks Stop?

## Why does accountability matter?

The question of accountability gets to the heart and the life-blood of the governing of schools.

In the Introduction we quoted a former Secretary of State, Kenneth Baker, in saying 'influence without responsibility leads to irresponsibility'. We see a second route to irresponsibility: responsibility without accountability. Fortunately, people in general have an intrinsic readiness to tell what they're doing or what they did, and to explain why they're doing it or why they did it. That for us is accountability:

- telling the what
- explaining the why
- to those who can reasonably expect to be satisfied.

The White Paper *Choice and Diversity* saw autonomy as the reason why accountability matters: 'The corollary of increased autonomy for schools is greater accountability by them to parents, employers and the wider community'.

A second reason why accountability matters is that schools, too, need the satisfaction and security of hearing the what and the why. Local employers, the wider community, the individual parents in the parent body, all form judgements about the school. What is more, they express those judgements, sometimes privately, sometimes publicly. It seems reasonable, then, to give them the opportunity to do something more than express them, and to expect them to be prepared to answer for their judgements, to explain, and to give the evidence. The Annual Parents' Meeting is the only occasion on which such *two-way* accountability might at present be exercised; disappointingly up to now, most meetings have been preoccupied with the one-way accountability of the governors' annual report. If the APM were expanded to cater also for wider community response, such a meeting could work at the exercise of mutual local responsibility (and hence mutual accountability) for the education and development of young people.

A third reason may be more compelling, as this poem by Fiona Norris reminds us all:

> They will not forgive us,
> These girls
> Sitting in serried rows
> Hungry for attention
> Like shelves of unread books,
> If we do not
> Make the world new for them,
> Teach them to walk
> Into the possibilities
> Of their own becoming,
> Confident of their exploring.

Accountability matters because young lives are at stake. That's why the stakeholders need to hear, and to be satisfied with, the account.

Such a view may come as a shock to the governor. 'Surely we can't be responsible and accountable for that?' Yes, the accountability is serious. A residential conference in Cambridge in 1991, which concentrated simply on the issues of accountability, ended with the encouraging view that people in education, from teachers and governors to officers and elected councillors, do understand both the nature of accountability and the fact that it is expressed in many forms. They know that they are accountable in many different directions and in many different ways. The various forms of accountability can be, and often are being, exercised with confidence. This variety may be implicit and hard to describe without thought, but it is not unclear. As Sir Michael Day, the then chairman of the Commission for Racial Equality, put it:

> Having different sets of accountabilities may leave you bewildered and uncertain, overwhelmed and not knowing where to turn or where duty lies. Yet it may also be the opportunity to respond in a way which exercises your own responsibility, and is therefore liberating.

On page 55 we offered three criteria that for us provide the touchstone principles of whether something is worthwhile in education. These principles, for us, justify committing the governing body to the exercise of accountability:

- accountability builds a greater whole – between the several parties who hear and speak with each other;
- accountability encourages giving responsibility – because those who exercise the responsibility are held on clear lines back to those who delegated it;
- accountability calls for a greater giving – in combining skills and wills for a common educational end.

The key practical issue is to find the connections between those giving the account and those receiving the account, and between those seeking the explanation and those providing it.

## Who's in the net of accountability?

Our definition of partnership sets high expectations: 'A working relationship that is characterised by a shared sense of purpose, mutual respect, and the willingness to negotiate. This implies a sharing of information, responsibility, skills, decision-making and accountability.' The image of the mutually accountable partnership that we want to offer in Figure 5.1 is of a net.

Any image has limitations. Three of the advantages of this image are, (1) in demonstrating the connectedness, eg, pulling in one place will twist the shape of the whole; (2) in emphasizing the cross-over points, the links in the mesh, the potential meeting-points; (3) in the ideas that one can generate around the word 'net', for example, of unravelling, of extending, of a finer or a wider mesh for the pupils, of a safety net. Pat Petch, an experienced chair of governors with whom we have worked, defines the scope of her governing body as, 'the *network* of people involved in decisions which ensures that policies have the consent of all affected by and contributing to them'.

Government/DfE

Office for Standards in Education

Local Education Authority

Education Association
(under the 1993 Act)

Governing    Head    Staff    Parents
body

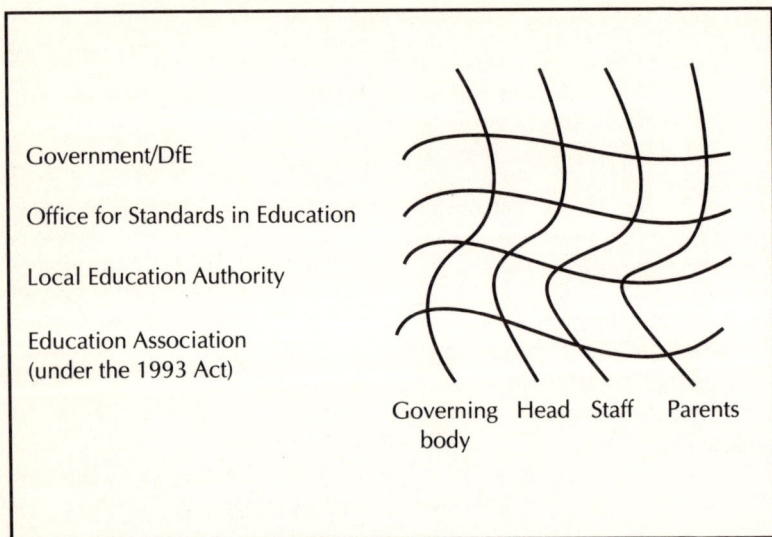

**Figure 5.1** *The net of accountability*

## The growth of the net

Perhaps the net looks simpler than it is. It used to be even more simple. The DES (as it then was) dealt with the LEA, the LEA dealt with the head:

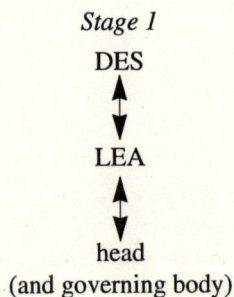

*Stage 1*

DES

⬍

LEA

⬍

head
(and governing body)

The 1977 Taylor Report prompted a much more significant profile for the governing body:

*Stage 2*

DES

↕

LEA

governing body  head

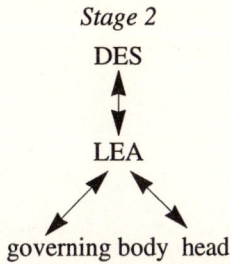

From the mid-1980s ministers and the DES judged it necessary, first, to deal direct with the head on some issues and direct with the governing body on others:

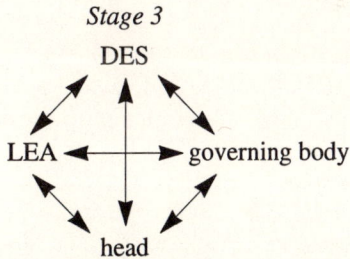

*Stage 3*

DES

LEA ←→ governing body

head

and, second, to give a higher profile to individual parents and the parent body:

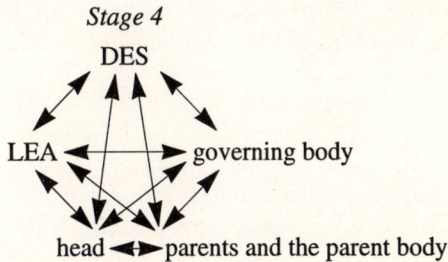

*Stage 4*

DES

LEA ←→ governing body

head ←→ parents and the parent body

Stages 3 and 4 were almost simultaneous. You will notice that the lines of contact have increased, from two in Stage 1 to ten in Stage 3. No wonder the net feels more complicated. Stage 5, under the 1993 Education Act, for a few schools will be the knitting-in of the Education Association. The Association temporarily will displace the governing body while a school at risk is 'put under new management until its performance has reached a satisfactory level', as the White Paper, *Choice and Diversity*, proposed. For those schools in the GM sector, the LEA has already been replaced by a direct line to the DfE (as it now is). Stage 6, for those GM schools and some other schools, will be the addition of the FAS which will allocate resources. At this point our image – and the real net – may collapse in an unusable tangle!

## Possible strains on the net

What the net doesn't show is the relative weighting, the pulling-power, the funding-power, the law-making power of the different potential partners; in reality there are strains. For instance, the government is part of the net, *and* has more power than anyone else. During the dispute about testing in 1993, the Secretary of State wrote to all chairs of governing bodies, urging them 'to make it clear that you expect the teachers in your school to fulfil their professional and contracted duty to administer the tests'. NAGM responded: 'In our view this is bad advice. We believe that governors will rightly resent any Government attempt to use them as an instrument for coercing teachers in a political fight between it and the teacher unions'. The net was being skewed.

A second issue is that of objectivity and 'interest'. In commenting on LEA and governing body publications at the time of ballots on GM status, the Junior Minister for Schools, Eric Forth, wrote: 'We expect governors to strive for accuracy in presenting their arguments, but we recognise that they are not in a position always to achieve that degree of objectivity and balance which is nevertheless to be expected of LEAs'. This suggests different rules for different partners. What about the balance and objectivity to be expected from the government?

A third issue is about rivalry, in this case, the absence of it. In its 1993 report, the Audit Commission noted: 'Most governors do not want to take over the Headteacher's role as manager of the school'. That's encouraging, and makes for a balanced net. However, the head who said, 'Let's face it! We're all in the business of conning our governing bodies' is definitely out to skew the net.

Another issue is that of influence. Much influence is exercised *outside* the lines of accountability, *between* the links, eg, through MPs, to the government; locally by clusters of governing bodies, to the LEA; by the Grant Maintained Schools Centre to the DfE. The heads have professional associations; the LEAs meet together; there is the National Confederation of Parent Teacher Associations (NCPTA); there are the governors' organizations, but none of them has legal status in the net of accountability. That's still to play for, turning influence into responsibility and accountability.

We think the government is right to see the interrelatedness of the school and its local communities, right to see that there are individual parents *and* the parent body as an entity, right to want a more immediate local feel in the governing of schools. The 1977 Taylor Report, *A New Partnership for Our Schools*, put it this way: 'We believe that all the parties concerned for a school's success ... should be brought together so that they can discuss, debate and justify the proposals which any one of them may seek to implement'. That emphasis on all the parties being assembled and meeting together suggests that the DfE and the FAS will need to be closer to each governing body than they presently seem to intend.

## The strand of professionalism

The government has successively challenged the power of the teachers' associations, the place of local politicians and the work of administrators, officers and academics. The government has sought to reduce their standing in education. It seems that it is suspicious of their expertise, their professionalism. At the same time it seeks to make each individual professional more accountable, through appraisal. This ambivalence raises problems for the governing of schools.

'Teachers are professional people, in that they maintain a relationship with a client (the pupil) in which the interests of the client are paramount; they have knowledge, experience and qualifications which they place at the client's service and they exercise individual judgement in the solution of a client's problems. They act in accordance with a set of values, so that professional conduct is ethical conduct. They are self-managing, accept responsibility for their own quality, and keep themselves up to date. In short, they exercise authority within their sphere of service. They take all these responsibilities, not only singly but corporately in institutions – the schools.'

This view, from Professor Tyrrell Burgess in a recent book, *Accountability in Schools,* expresses one of the ways in which professionalism and our attitude towards it is a key issue in the governing of schools.

Two associated key issues for the governing body are:

- how to get the professional views of the teacher-governors fully contributing to the governors' discussion;
- how to help the head to come to see and think as a governor. The head, even when nominally a governor, cannot lay aside the unique role of

head, as professional lead-officer and chief-executive for the governing body. The two roles cannot be exercised simultaneously. Heads need to be conscious of which role they are in.

## A professionalism proper to governors?

Perhaps an even more important issue is how the individual governor can find a way to the proper perspective of a governor, what we want to call 'the professionalism of the governor'. And will the government reject the governors' expertise, their acquired perspectives and points of view if they do? Perhaps it will, with the disclaimer: 'They've gone native. They've joined the educational establishment'. The government cannot have it both ways: if governing bodies are to do a real local job, which the government seems to want, the governors will acquire a distinctive local view and national voice. The government cannot ask for the accountable exercise of local responsibility, as it is doing – and, we believe, rightly – and then dismiss as corrupted experts the very agents of that accountability.

Christine Webb, a teacher at that conference in Cambridge about accountability to which we referred earlier, spoke of

the lovely story of a bright young adviser visiting a school and enquiring of the tough, wiry deputy head, 'To whom do you hold yourself accountable?' She fixes him with her glittering eye, and replies, 'The children, and God'. She went on, 'Some of us will have to replace God in that answer with a more fallible authority, but the equation is essentially the same: honest endeavour towards the pupils, plus honest transmission of the truth that is within you, equals (you hope) enough peace of mind to sleep in your bed at night. At any rate, those are the standards to which we all turn when priorities seriously clash. No legal framework, no formal map, no economically viable flow chart can support you if you feel you are failing in those central truths.

Something of the same, 'the standards to which we all turn when priorities seriously clash', is what each governor and governing body needs to find as its base line, its professionalism, for its exercise of accountability.

## What do we mean, 'accountable'?

We hope in this section to give you enough from which to form or check your own sense of accountability. The accountability conference in

Cambridge in 1991 grew into a book, *Accountability in Schools*; we recommend it, and our quotations in *this* section, unless acknowledged otherwise, come from it.

*The Independent* in July 1993 commented that:

> The Government and the teaching profession will need to reach agreement on the central question of public accountability. John Patten, Secretary of State for Education, has argued that league tables ensure that teachers are accountable to parents. But teachers say the existence of the National Curriculum, Governing Bodies, Standardised Tests, and regular inspections already provide that accountability.

Evidently, using the same word isn't enough ground for agreement.

At the beginning of this chapter we said that for us accountability is about telling the what and explaining the why to those who can reasonably expect to be satisfied. Another way of putting it is that accountability is about 'the line between':

- the line is of explanation
- the line is between people
- the people each have responsibility for certain things
- they accept the responsibility whether things go right or wrong
- the explanation is about what and why
- the explanation is also about other questions
- the explanation is needed until the person hearing it is satisfied.

'Autonomy' doesn't come into it. Autonomy is precisely *not* having to explain to anyone. It's a misused and misleading word in the governing of education. What governing bodies need is the freedom to get on with things, with full enough delegation and funds. Being accountable means a readiness to answer, as the price for having had the freedom and responsibility.

## The line between people

The line of accountability is between people. Tyrrell Burgess suggests where it begins:

> Parents are a child's first educators. They have a natural and legal obligation to protect them and not to do them damage. In education law they must see that their children are educated. If parents choose to educate their children at home, rather than at school, the (local) authority must

93

decide whether the education is full-time, efficient and suitable. In other words, parents are legally accountable to the authority (it can call them to account) for their duty to educate. It is unlikely that this thought ever crosses the minds of most parents. They feel accountable, but the accountability they feel is not legal but personal. Some parents may still have a lively idea of accountability to God, and a few may believe that they will render this account at the last trump. Most, however, have a substitute: they feel accountable to themselves, to their own consciences, to a set of moral values, to public opinion – represented by friends, other parents, a social circle – which creates shared expectations. Of these two kinds of accountability, it is interesting to note which is the more effective. The legal accountability is largely unknown: when parents fail, the legal sanctions do not work. It is not the law, the formal accountability, that makes parents talk to teachers, go to parents' evenings, worry about a choice of school or agonize over GCSE options. It is a parent's sense of accountability to what is personally and socially thought fitting.

This personal moral accountability exists alongside other person-to-person accountabilities. For instance, for the parent there's one line with the teacher in school; there's a different line with the private piano teacher contracted to give opportunities at home to the child. As we see it, a contractual accountability has its place in education, but it is not enough for education. Del Goddard helps to make the distinctions:

> Confidence and quality are more likely when contractual accountability is the safety net, professional accountability the support, and moral accountability the driving force. Together they make the structure. Together they provide the means for dealing with the unacceptable, for solving problems, for developing teachers and schools, and restoring the confidence and morale of the service and the public. Each accountability on its own is inadequate.

The governing body has three important person-to-person lines relating to it: to the Secretary of State, on behalf of the government and the nation; to the parent body on behalf of the children; and to the LEA (in most cases) on behalf of the local community. These accountabilities are neither contractual nor professional. They seem to add up to what Angela Thody calls 'a new focus of democracy'. The governing body has responsibility for certain things (which we sketch in the next section). The Secretary of State, the parent body, and the local community also have responsibilities in the line that relates each of them to the governing body. For instance, the Minister has the responsibility to give the governing body the powers and the resources appropriate for its tasks and responsibilities. For parents, Kenneth Baker, a former Secretary of State, expresses the responsibility sharply:

### 'A joint responsibility'

... Whatever reforms I am able to bring to the education system, they will not produce the results we all want to see unless children are able to benefit from schooling. Whether they are able to benefit is something determined more by the attitudes of the home than by the attitudes of the school. We rightly look to teachers to teach, but it is the joint responsibility of parents and teachers to educate.... We often talk about the partnership in education. Of course parents have a right to expect schools to provide good education, and that is why we are undertaking radical reforms of the education system. But perhaps we lay insufficient stress on the responsibilities of parents in that partnership. Teaching is a difficult enough task made even more difficult when parents don't take their responsibilities seriously enough.

As for the local community, it has the responsibility to engage with the governing body, its representatives.

## The line of explanation

We suggested earlier that accountability is essentially about what and why, and that explanation is needed until the person hearing it is satisfied. For the governing body the explanations need to be about their thinking, their decisions, and their direction for the school. The governors' annual report gives part of the what – that part which deals with what has been done. The 1993 Act will require governing bodies to add the why with regard to any decision not to go for GM status. But there is also the what and the why of current issues and of expected future problems and plans. The parent body is interested in having a say in the life and learning of the school, today's issues and tomorrow's issues, so research for the DfE has established. The governors' annual report is welcome, but the report is the past and the child – and the parent's interests – are present and future.

Explanation requires people to talk and to listen. Susan Heightman, as a parent, is clear that: 'Parents and teachers both need to give an account of their perception of the child to each other, rather than demand explanations or have recriminations'. Teachers need to talk with, not simply report to, individual parents, and the governing body needs to talk with, not simply report to, the parent body. The parent body needs to be satisfied; at present many parents do not believe that schools are listening to them. If the governing body can see the parent body as part of the school, *inside* the school, then the handling of the accountability, even handling any mistakes, is less threatening. Getting it wrong is manageable if head and governors admit responsibility, express regret and recognize the context which allows them to try again. David Hart, general secretary of the National Association of Head Teachers, saw it too narrowly when he wrote: 'Heads are accountable because they can be subject to disciplinary procedures.' That's true, but it simply expresses the contractual element of accountability, not the moral or professional elements.

## Responsibility, right or wrong

Among the governing bodies in 1993 which were struggling with what the government required of them, one took a determined stance: 'Our primary school governors have refused to do the Government's dirty work in sending school performance tables to all parents. Backed by the Nuremberg principle that immoral orders should be disobeyed, we ignored the demand to distribute. True, it may mean breaking the law'. That governing body found its own criteria of right and wrong, accepted that its view was contrary to the law, and recognized that it might leave itself open to accusation and recrimination. The point we are emphasizing is the readiness to live with the consequences of the governing body's own decision, not necessarily the rightness of the decision.

Certainly, accountability leaves head and governors vulnerable to penalty. The price or penalty that might be paid must be proportionate to the authority and powers which the head and governors have at hand to exercise, and proportionate to the consequences of what has been done or not done. Pat Petch, as governor and chairman, finds that:

> It is very easy to demand accountability, to write documents promising accountability, even to draft legislation requiring accountability. It is not so easy to achieve effective accountability in practice. Success or failure depends on attitudes. If people feel that what is required is fair, that it is part of a two-way process which recognises their views and values, that it is worth-while, that the aim is to improve the service rather than threaten the individual, then they will be prepared to answer questions, share information and discuss what they are doing and why and how they are doing it. They will be prepared to help establish processes which lead to

accountability and then participate in ways which ensure it is genuine. If people feel threatened, if they feel that their views do not matter or that it is not worth-while, any process to make them accountable will be a token. Successful accountability can enhance the quality of relationships: token accountability may damage them, as people become resigned, disappointed or hostile.

## Where to begin

For this 'new focus of democracy', heads and governors need to imagine, create and develop the appropriate ways of holding the line between. If the machinery isn't there, we are all free to make it for our own local and national needs. The place to begin, for each of us as governor or head, is with our personal self. Table 5.1 suggests the key points.

**Table 5.1** *A governor and head's declaration of accountability*

> - My responsibility is to be a governor, to advance the life and learning of the children.
> - I am responsible for what I choose (thoughtfully?) and what I assent to (unthinkingly?).
> - I am responsible for what I do, for what I don't do, for the tasks I seek or accept and for those I ignore or reject.
>
> I am accountable to myself first, and to my governing body second.

## What's the governing body in particular accountable for?

In focusing on the governing body's accountability, we suggest you remember the net we sketched in an earlier section of this chapter (page 88), to make the point that little is solo, most is shared. We suggest that there are six dimensions of the governing body's commitment:

- advising
- policy making and development planning
- coordinating and mediating
- supporting and promoting
- monitoring
- improving their own working methods.

This adds up to governance rather than management, directing rather than operating. The view we are advancing is well put by Tyrrell Burgess, in

*Accountability in Schools*, that '... governors act best when they see themselves as the means through which a school is accountable. It is a semi-political accountability, in that they are required publicly to justify themselves to, but are not revocable by, the public'. Or, as Felicity Taylor of the Institute of School and College Governors puts it, the governing body is the vehicle of accountability. That means that the governing body has some accountability for all that happens in the school, all that the school is, all the school is not and has yet to become. No wonder the responsibility feels daunting at times; no wonder heads and governors are nervous of Annual Parents' Meetings.

The running of the school is the head's responsibility; the accountability is *to* the governing body. The governing body is accountable *for* the head – and thus, at one remove, for the head's actions and omissions. To quote Tyrrell Burgess again:

> If a Headteacher is accountable to the governors, accountability exists: if the governors usurp the Headteacher's functions, they become largely unaccountable. The point was made starkly by the experience of the grant-maintained Stratford School: the governors sought to usurp the functions of the Head, to exercise their responsibilities directly and not through professionals – and they became at once out of control, even of the Secretary of State.

Together, the governing body and the head (whether or not a governor) have certain responsibilities and tasks:

- to lead and develop all those in the school community
- to take risks for quality
- to engage with the community and its values
- to tune the staff and the governing body to a wavelength of mutual understanding.

Together they are accountable.

The head as head, and regardless or whether or not a governor, has distinct responsibilities and tasks:

- to engage with lay values
- to use and supervise the available powers
- to organize the running of the school and the service from the staff
- to implement the governing body's policies.

The governing body is itself responsible and accountable for developing the means to receive the head's account and to hold the head to account.

The governing body, including the head only if the head is a governor, has particular responsibilities for which it is accountable:

- to secure service from professionals
- to consent, to sanction appropriate proposals
- to mediate and co-ordinate powers
- to support
- to advise
- to direct
- to carry the accountability for policy and for those who operate it.

'But', we hear you cry, 'that doesn't tell me what to *do*!!' No. We're not writing that kind of book. We want you to have a frame of reference about governing, from which the head and governors can negotiate what needs doing, most, now, in your school, recognizing that it will change. As Chris Lowe, the head of Prince William School in Northamptonshire put it: 'I want to see less wallowing around in the morass of duties still coming thick and fast, and more concentration on pockets of expertise.... I would like to see knowledge and understanding of the role *lead* the skill of discharging it, rather than vice versa'.

A reminder about where heads and governing bodies actually are now, comes from the Audit Commission's report which we quoted earlier:

In some 40% of schools accountability is weak or non-existent.

Most governors do not want to take over the headteacher's role as manager of the school.

## Giving the account and being held to account

You probably remember the net we offered as a way of understanding the pulls and tensions within the education system. At the end of the first section of this chapter we used the image of the net again when we said that the key practical issue is to find the crossover points that connect:

| those giving the account | those receiving the account, |
|---|---|
| those seeking the explanation | those providing it. |

In this section we look at the governing body giving the account and providing the explanations to parents, to pupils and to others, to the present and to the future. The governing body's trusteeship for the future means that the account and the explanations can be used as a basis for development.

**Accountability is developmental,
just as staff appraisal is developmental.**

## Accountability inside the governing body

We pick out two dimensions of giving an account and being held to account. The first is within the governing body's own working. We list in the left-hand column of Table 5.2 some opportunities for being open to account, many of which are now widespread in most schools, a few of which may be quite new to you. In the right-hand column we suggest for each opportunity one important question that the governing body would do well to discuss. What we want is that you see the scope for handling accountability.

**Table 5.2** *Internal opportunities of accountability*

| | |
|---|---|
| * The termly report that monitors the budget | * Who drafts the report? |
| * The head's report to the governing body | * Who reviews the parameters? |
| * The school development plan | * Who discusses and drafts it? |
| * The Governors' annual report | * Who drafts it? |
| * The annual report to the governing body that monitors the success in practice of their policies | * Who does the monitoring? |
| * Annual self-appraisal by the governing body | * Who leads? |
| * Statement of values, aims and objectives | * Who reviews? |
| * The prospectus. The complaints procedure | * What do the parent-governors think of the paper's 'usefulness'? |
| * Appraisal and/or self-appraisal of individual governors. | * What are the pros and cons? |
| * 'Whistleblowing' on crises. Getting the message to parents, the press, LEA, or the government. | * Does the governing body know how and when? |

There are other practical and practicable contributions. When did your governing body last consider having photographs of governors in the entrance hall, a governors' notice board, a 'surgery', wearing 'Governor' badges at parents' functions, sending a newsletter to the interested parties represented in the governing body? Singly, each may seem trivial. Collectively such arrangements create the line between and encourage other people to come forward with their own agendas. The governing body's evident accessibility and responsiveness are part of its accountability, whether to the local journalist, the aggrieved group of parents or the parish council.

## Accountability to the world outside

As head or governor, you can probably identify procedures, occasions and mechanisms, that present the governing body's accountability to the world outside; here are a few:

● sharing the budget monitoring, *with the LEA*;
● publishing the head's termly report, *to the parents*;
● distributing widely the governors' annual report, *to the community*;
● displaying the school's development plan in the entrance, *for all to appreciate*;
● encouraging the use of the complaints procedure, *to get worries into the open*;
● the four-yearly inspection under OFSTED's supervision;
● the Annual Parents' Meeting.

It's important to remember that, for the sake of the children, those holding the governing body to account have teeth. The LEA may recall some (or all) delegated powers, the registered inspector may declare the school to be 'at risk', the governing body may be replaced by an Education Association, and there is some legal and contractual accountability in extreme cases. Remembering these does not mean letting them paralyse the governing body.

## Weak mechanisms of accountabiliity

We referred to the registered inspector. The inspector's report on the school is, in a very real sense, a report on the governing body's discharge of its accountabilities. Part of the report will be about the direct contributions and omissions of the governing body. The government, reasonably, could not delegate so much authority, with the necessary powers, and *not* see how it was going. Whether the particular arrangements, under the Schools Act 1992, are appropriate is another matter; they do not seem to be developmental.

There are two ways of conducting an inspection. The first is to find answers to a series of questions: What does the school seek to do? Is this worth doing? Is it done well? It enhances accountability by giving an independent view of the school's own account of itself. The second method is to arrive at the school with a checklist and tick off the extent to which the school matches it. It is an instrument, not so much of accountability, as of control. It replaces independent judgement about what a school is trying to do with pressure on the school to do what someone else wants.

Notwithstanding this criticism from Tyrrell Burgess, the governing body's focus in discussing the specification for the inspection, the report and its action plan must be: 'How can we secure more light, time and energy to better discharge our accountabilities for the children? We'll recognize and admit our limitations. We want active partners with whom to build'. The lay inspector is someone with whom the governors may strike a common chord.

In a previous section we referred to the government's interest in league tables. Not only league tables, but Parents' Charters, parliamentary questions, parents at MPs' surgeries, the chief constable, the industrial tribunal – all are notional contributory roles and processes of external accountability. What the government hasn't found the means to develop are (1) the readiness of the governing body to be accountable, and, (2) the self-confidence of the parent body that accountability is, in part, theirs. The response of the NCPTA to the White Paper, *Choice and Diversity* included an argument for parents to have a right to form a home-school association at each school. The important question for us is whether parents are ready to carry their share of accountability.

## 'The Annual Parents' Meeting – Chore or Challenge?'

Following some research in 1992, the DfE published a leaflet, *The APM – Chore or Challenge?* for governing bodies; a copy was available for every school. The research set out to see why APMs were not working and what factors were helpful if and where they were working. What the research found was that, rather than being an occasion for the governing body to give an account and to be held to account, the APM had been swamped by the governors' annual report. The reports have improved, but they are retrospective and there is not much to discuss about the past and, as Joan Sallis, founding chair of AGIT, has commented, 'There is no participation in the past'. Parents certainly want the APM, as an opportunity to have a say; they don't want the *only* say. Parents feel that the head and governors have not been ready to listen to views and to discuss issues about the present and the immediate future of the school. Given the priorities for heads and governors – of LMS, the National Curriculum, assessment and the need to get some assurance that their past efforts have been (at least) acceptable – that limited listening is understandable. But the focus on the past is why parents have not come. The APM has typically been:

| THE SCHOOL | | the parent body which |
|---|---|---|
| governing body | telling, informing | listens and endorses |
| head and staff | | what's gone on. |

In a fair number of schools, on one or two issues at a time, there has been another mode of dialogue:

| THE SCHOOL | the parent body, which |
|---|---|
| sharing and enquiring | discusses, then responds |
| with | – and influences the |
| | school |

This seems a more developed form of partnership and accountability, with the governing body perhaps saying, '*We're* responsible, but what do *you* think we might do?' This seem likely to be the next phase for more schools. In a very few schools, the line between has broadened further with:

SCHOOL
governing body
head and staff

negotiating with

the parent
body

on terms of joint responsibility in
developing and framing policies.

The governing body remains the accountable partner, but it has incorporated the parent body into the school, into ownership of the school and its life and learning.

## Whose is the APM?

This raises the question of whose is the APM? The governing body is responsible for its convening and its conduct. Generally, influenced by the government, it has itself conducted the APM as an annual general meeting – which is a format which offers a very limited form of accountability in most companies! However, there is an argument that the parent body itself might have responsibility or joint responsibility for the APM in its agenda and its conduct.

As a mechanism, invented in 1986, the APM has suffered from neglect by the DES/DfE, from other higher priorities for the head and governors, and from being dependent upon a number of sophisticated understandings which several different, busy groups of people need to establish to make the APM work well. It's a very good example of the fact that accountability has to be worked at.

As preliminaries, the parent has to be satisfied as an individual parent, and the parent has to see that the parent body works for all children. The head has to understand the role of head as the leader of the staff team and with an overview similar to that of the governing body. The governors have to understand the role of being an individual governor. The head and governors all have to understand their responsibility, jointly and separately as head and as governing body. All have to understand that 'communication' and 'partnership' are not just 'giving information'. All have to understand that legality does not have to mean formality; that the APM is a relationship, not an agenda; that the working language of the governing body is not the parent's language in discussion; and that the head, staff and governors have to foster open-ended discussion in small groups with parents.

All this understanding comes before the parents, the governors and the head start to understand the place in the system and the power and the authority of the APM. The same is true for the LEA and its responsibility under the 1986 (No 2) Education Act, which most ignore. Then they need an understanding about collaboration and accountability. No wonder most APMs have not got beyond first base in their first five years.

Meanwhile, one simple way in which any governing body can advance is just to plan the stages around three termly meetings of the governing body, as shown in Figure 5.2. Accountability can be made recognizable by finding procedures and occasions that join people together.

## 'A bike or a bollocking?'

Accountability is the line between, to and from. It's checking, genuinely, that the job has been done and asking, openly, 'How can we do more for the children?' And it needs to be understandable.

A father rings Kent County Council for advice. His son has achieved Level 1 at Key Stage 1. 'What does it mean?', he asks.

A council education officer starts explaining the 10-level scale and the key stages of the National Curriculum but the father interrupts.

'I don't want to know all that' he says. 'All I want to know is, is it a bike or a bollocking?'

It is a funny – though sad – true story. Here is a boy with some serious problems, but because of the alien reporting system to parents, his dad does not know whether to reward him or tell him off.'

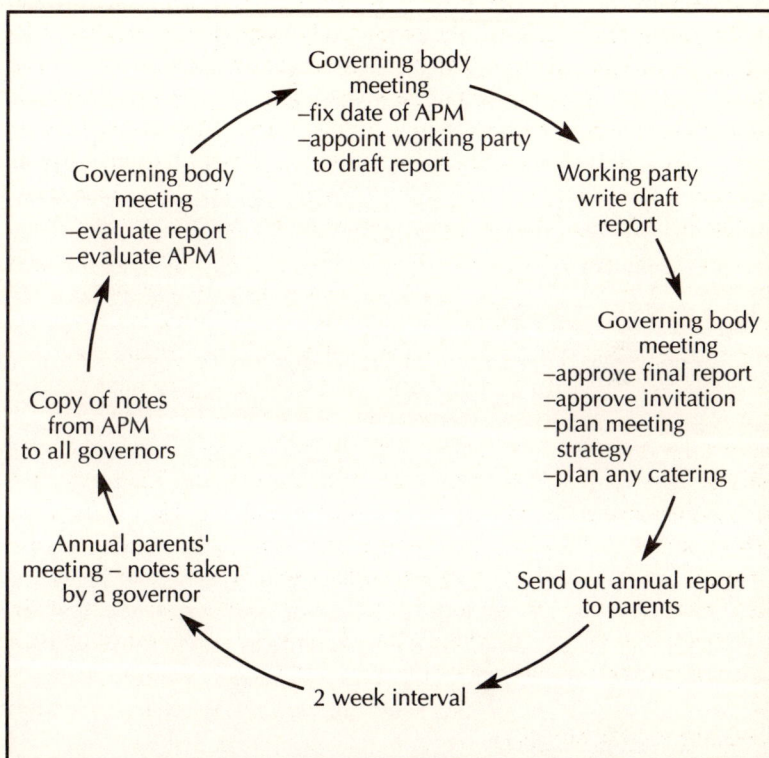

**Figure 5.2** *Planning the stages for the APM*

Those who are serious about the principle of accountability will do more than go through the motions, will do more than offer the set speech. They will check to see that the message has been received and understood.

## Receiving an account and holding others to account

For the governing body, it's probably no easier to hold others to account than it is to go forward to present an account. The government model for the exercise of accountability doesn't offer much help. It seems to see general elections as its only measure and process of accountability. It sees no real line of developmental accountability between itself and those it needs as partners.

The model of parent and child may be appropriate for the governing body in its degree of openness to the head and the staff body when holding them to account. A parent puts it this way: 'I know nothing of any significance about physics, but that does not stop me from asking Melissa what she has learnt'. So the child tells the story of her learning; the child becomes the teacher of the parent and grows in self-esteem. There is no need for parents to know everything. Children need to see that their parents are still learning too and still asking questions. In the same way, there is no need for the governing body to know everything. But, in holding the head and staff to account, the governing body will need to ask questions; the art is to focus on the answers that are relevant to the aims and quality of the school, rather than on simply more knowledge for the governors, or on identifying limitations in the head and staff.

## Reporting

The governing body could regard the head's termly report not simply as a briefing paper or an historical record, but also as a regular report about the discharge of responsibilities. The range and timing of regular items and occasional special reports might be better spread between the head and certain governors, for example chairs of committees or the governor with responsibility for health and safety. The Audit Commission's advice in 1993 was that the governing body should receive at least a termly report on expenditure and the budget.

## Appraisal

The governing body should regard the appraisal of the head as their concern; the head's accountability is to the governing body. The process properly needs to harness professional criteria and assessors, but the governors' evidence, their judgements, their discussion and their conclusions are a necessary part of fair appraisal of the head as their chief executive. The connection with performance-related pay and the annual review of salary is sensitive. The money is substantial. The use of the money for the head's salary rather than for that of other staff or for other needs in the school may be right. It needs open discussion, not an embarrassed near-silence with the aim of keeping the head happy. Above all, the performance needs to be related to objectives.

## Procedures protect

Grievance procedures are the healthy way of surfacing discontent, for staff and the head. Complaints procedures offer the same healthy channel for parents; and disciplinary procedures are the healthy way, clear of personal embarrassment, for recording valid disapproval. It is not a failure if the procedures are used. There is much more likely to be a concealed and festering problem if the feeling prevails that, 'we don't want to be formal about it'. Formality is precisely the way to reduce the emotional confusion. The governing body needs to attend to, publicize and encourage the use of procedures, rather than allowing the embarrassments, tangles and ill-will that usually go with informal, ad hoc manoeuvres, however well-meaning. A proper use of procedures protects and ought to ensure the best possible outcomes.

The registered inspector and team call by but once in four years. The governing body needs to develop, with staff, effective arrangements for regular reviews of school practice and the reality (or otherwise) of agreed policies. The governing body needs to receive and consider an annual monitoring report.

We summarize these practicable steps in accountability as follows:

- The discharge of responsibilities
  - included in the head's termly report,
  - regular reports from governors with particular responsibilities.
- The head's appraisal
  - evidence,
  - discussion.
- Performance-related pay
  - objectives,
  - discussion.
- Using procedures for dissatisfaction.
- An annual monitoring report on quality achieved.

It is difficult. A 1993 survey in Warwickshire found that, 'it was felt that by far the most important function of the Governing Body is to support the work of the school while its role as critical friend or mediator is not experienced to be very significant'. But as the County Education Officer observed, 'Being supportive and encouraging, and being accountable and business-like, aren't mutually exclusive'.

We quoted earlier Kenneth Baker's comments about the responsibility of parents. The home-school contract is one way to make explicit the mutual accountabilities; the NAHT developed one effective, widely-used model. The response to the four-yearly inspection report is another way to involve the parents, harnessing their help in preparing and implementing the action plan. Pressing parents for their views is another way – and then taking them seriously. A head reported this conversation with a parent:

> 'Why do you think it is a good school?' he asked with interest. As reluctant as I suspect many Heads are to embark on definitions of the good school, I turned the tables and suggested that, as a parent, he should have ideas about the good school. He did. He replied immediately and carefully that he knew exactly the kind of school that he wanted for his child: 'a school that offered a much broader curriculum than he had received at his school, a school that encouraged a sense of responsibility for society, and a school in which his child was educated alongside the broadest possible range of pupils from all backgrounds'. [The head went on] If a parent can be that clear in defining the school he wants, then a Head can be equally clear and indeed courageous in defining the school's aim.

## Accountability is more than OK

We hope we have convinced you that accountability is thinkable, is practicable and needs procedures. We believe that the education system is benefiting already from 300,000 responsible governors who are putting to good use their better knowledge and understanding of the way schools work.

## References

86. 'They will not forgive us', Fiona Norris in *Yesterday, Today, Tomorrow,* eds Leggett and Moger, ILEA, 1987
87. Michael Day, in *Accountability in Schools* (reference 9, above)
90. Walter Ulrich, Information Officer for NAGM, *TES,* June 1993
90. Audit Commission, *Adding Up The Sums,* 1993
94. Angela Thody (reference 73, above)
95. A Joint Responsibility (reference 29, above)
98. Felicity Taylor, letter in *TES*, 1993
99. Audit Commission (reference 90, above)
104. 'A bike or a bollocking?', Roy Pryke, *TES*, 1993
105. Figure 5.2 'Planning the stages for the APM', Ann Cains, in *Managing Schools Today*, November 1992
107. Survey in Warwickshire (reference 67, above)
108. 'Why do you think it is a good school?', Maggie Pringle (reference 9, above)

# Section 2

# A Pulling Together

# Chapter 6

# Getting the Picture Together

## Our values as yardsticks

We see this chapter as the essence of the book. In it we draw together the analysis and suggestions of the previous chapters on five different facets of governing:

- Quality
- Politics
- Imagery
- Ownership
- Accountability.

We relate these five facets to the government's five key imperatives:

- Quality
- Diversity
- Choice

- Accountability
- Autonomy.

In Figure 6.1 we also highlight differences between six important ideas that easily get confused and cause confusion.

**Figure 6.1** *Don't confuse . . .*

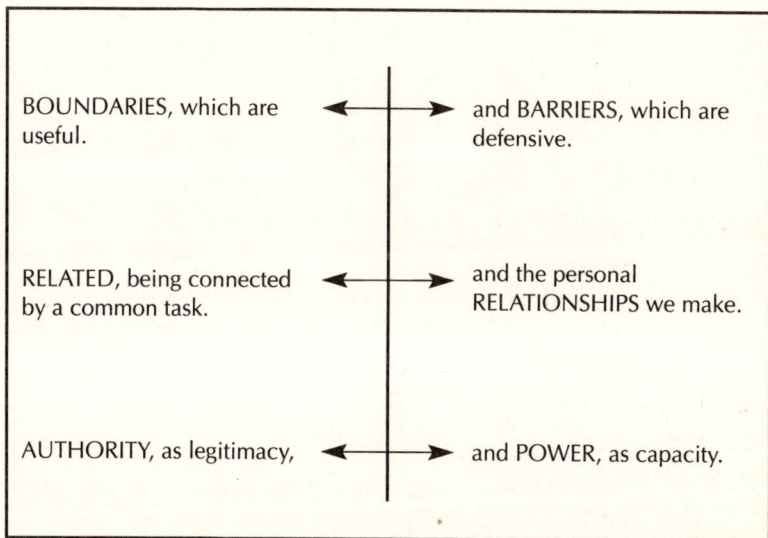

| | |
|---|---|
| BOUNDARIES, which are useful. ◄——————► | and BARRIERS, which are defensive. |
| RELATED, being connected by a common task. ◄——————► | and the personal RELATIONSHIPS we make. |
| AUTHORITY, as legitimacy, ◄——————► | and POWER, as capacity. |

We hope this chapter will be a tool for you as governor or as head to put a picture together in your own mind of what governing really is all about. That picture will give you the self-confidence to prioritize your work through the governing body on the basis of a set of principles, rather than in deference to the workload pushed at you by others. We want governors and heads to get clear of feeling like a dustbin which Helen Goff has felt.

### ON ME ... AS A DUSTBIN

Sometimes I feel like a dustbin
Filled and ever filling
With things disposable
Junk food
Polluted air
Adults' opinions

And I say
**'STOP, STOP**
You're filling me
Too full!'
And they say
**'TOUGH'**
And teach me
**TRIGONOMETRY.**

I want to know what happens
When I flip my lid.

## What would you fight for?

For ourselves we have not found a better way than the Taylor Report to express what's needed (the emphasis is ours):

> We believe that *all the parties* concerned for a school's success – the LEA, the staff, the parents, the local community – should be brought together so that they can *discuss, debate and justify* the proposals which any one of them may seek to implement. We consider it necessary that all the parties should *share in making decisions on the organisation and running of the school* since, in our view, this is the best way of ensuring that every aspect of the life and work of the school comes within the purview of all the interests acting together.

The Taylor philosophy seems to answer 'Yes' to our three criteria for good governing, ie, whether a particular development:

- advances a greater whole in education
- increases and widens responsibility
- calls for more practicable generosity and commitment.

The present arrangements, which include the existence of GM schools, suggest the need to extend the Taylor recommendations to include the 'real presence' of the government in the partnership as 'a concerned party'. We have alluded already to the fact that if government is determined to be involved at this level of detail in education then the gap between it and the other partners is too great. In a new system that is evolving, this and future governments will have to decide to create means for closer engagement or to back further away from the detail of strategy and implementation while still remaining concerned with policy and objectives.

The fundamental question about quality may be, 'what would you fight for?'

Fighting for being 'at one with myself' is what we're recommending to you as head, you as governor, you and your colleagues as the governing body. It will determine the quality experienced in the school, the integrity of the school.

## Keeping it simple

We're recommending, too, that you keep your quest for quality simple. Robert Fulghum put it this way:

### All I really need to know

All I really need to know, about how to live and what to do and how to be, I learned in kindergarten. Wisdom was not at the top of the graduate-school mountain, but there in the sandpile at Sunday School. These are the things I learned; share everything, play fair, don't hit people, put things back where you found them, clean up your own mess, don't take things that aren't yours, say you're sorry when you hurt somebody, wash your hands before you eat, live a balanced life, stick together, goldfish and hamsters and white mice all die – so do we.

In Chapter 1 we argued that values are the core of a child's education, and that it's in the head's and governors' discussions of their own values and the values of others that the governing body will find the grounds for agreement on what it means by quality. 'We haven't got time for such abstract discussions,' commented one governor. Yet without such discussions we see no way for the governing body to fulfil its responsibilities to provide a curriculum statement, give guidance on discipline and ethos, and produce a statement of school values. All of these legal requirements provide the basis for subsequently determining priorities and making decisions. For instance, if a school values *all* pupils equally, then its spending on special needs and on equal opportunities will reflect that value.

## Focusing on quality

We remind you that in *Choice and Diversity*, the government provided us with a relatively simple statement of its values by listing five key imperatives – quality, choice, diversity, autonomy and accountability. In pursuit of its imperatives the government can set standards and it can inspect what is achieved, but it essentially needs the governing body to make sure that standards are achieved and, if possible, rise. There are certain things, there-fore, which the governing body, in its essence, is *not*.

● A rubber stamp
● A delivery system
● A local agent
● A big power-base
● A company branch.

The governing body is, in its essence, a locally-generated, responsible body. It is locally accountable for the process of quality:

● it takes stock
● it sets goals
● it takes and supports action.

The governing body's focus is the whole school. Yes, the school is multi-layered, even complex, but for the governing body to narrow its attention to only a few dimensions will stunt and warp the growth of the school, of the children's life and learning. You may like to look back to Figure 1.1 (page 14) and our suggestions about the wide range of facets of the school, and to Figure 1.2 (page 15) to consider the characteristics of effective schools. We suggest four focuses in your action aimed at quality:

● A development committee or working group.
● A development plan or papers.
● The Annual Parents' Meeting.
● The self-appraisal of the governing body.

These are practicable means of concentrating on the raising of standards and achieving quality.

We encourage you to focus your judgements on your own performance on relevant evidence about the extent and value of the governing body's contribution to the school; the format shown in Table 6.1 may prove useful.

When you have decided what you value, those values need articulating, in both senses of the word. You need to discover the connections and dis-connections between different sets of values, eg, staff values on learning by experience and parents' values on learning the rules. The governing body is the place to focus such articulation. Values need expressing, often as feel-ings, like the expressing of the head's disappointment that the parent body doesn't discuss or comment on the curriculum, and the parent body's feel-ing that the head and governors are too busy to listen.

**Table 6.1** *Self-appraisal for the governing body*

| How much do we think we are contributing? | | | |
|---|---|---|---|
| Little? | Something? | Useful enough? | A great deal? |
| How well do we think we're doing? | | | |
| Poor? | Fair? | Well enough? | Very well? |

## Living with uncertainty

Finally, if quality is developmental, can you live and govern in some uncertainty? Can you live with and work constructively through the difficulty and the pain of problems?

- Can you avoid deferring the difficulty? *Can you tackle it now*, as best you can, and get it behind you?
- *Can you accept your responsibility* – for why things are the way they are; for trying to improve things; for the consequences (good, mixed, bad, foreseen and unforeseen) which will come after you've tackled the difficulty and made *a* solution for it?
- *Can you go on trying to understand* what governing is about?
- *Can you accept the need to find a balance*, rather than to find perfection – between conflicting needs, goals, duties and responsibilities? The governing body as juggler rather than conjurer? You may not be able to pull the white rabbit out of the hat today, but you can try to keep all the balls in the air.

## The role and the task in *my* mind

This section picks up the government's key theme of diversity and our earlier chapter on imagery.

### Talking to yourself

Talking to yourself is one way of working things out: 'My problem is ...' We offer some prompts of what you might keep saying to yourself:

● Be myself. Stay myself. Give myself. Risk myself.

- There isn't a governor mould into which I must pour, trim, stretch myself.
- Get a view of the task of the governing body
- Get a view of my contribution to that task. Whatever it is ... that view is my role, me being a governor, this year, in this governing body, for these children.
- See who and what are the other parts that have a connection to me – where I fit in the system.

Another way is to picture it to yourself, or to map it out. In a training seminar, we asked secondary heads to draw diagrams which related themselves to the rest of the organization of the school, including the governing body. One head stood back from his diagram. It was neat and well ordered – until it came to the governing body when, frankly, it was a mess. 'I don't understand it', he declared, 'I don't know what they are supposed to *do*!' 'Give me the *verbs*', he shouted. We brainstormed some verbs, but we didn't give him a blueprint; it depended on his governing body and his school and their collective view. For you, it depends on what suits you and your governing body. These are some possible personal contributions, two or three of which may particularly suit you at any one time.

| | |
|---|---|
| Shape | Chair |
| Finish | Help |
| Mediate | Negotiate |
| Organize | Lead |
| Plan | Administer |
| Decide | Advocate |
| Specialize | Evaluate |
| Build the team | |

As you look around and identify your (potential?) contribution, then you've found *your* role in the governing body. The other individual governors will have different views about the school and the task of the governing body, different personal views for their role. 'Why aren't we proactive as well as reactive?', may be one fellow governor's preoccupation. Each of you and your views are connected by your membership of the governing body. You may have uppermost in your mind, 'How can I get my contribution heard? There's so much I want to ask. They're so impatient, and won't give me space to talk'. (For help on this see Chapter 7.) Your diverse views and skills must be brought together.

The two key elements of diversity which we are emphasizing are that 300,000 persons are not to be moulded, as model governors; 25,000 governing bodies are to be shaped, by 'How we see the task in this school'. In each school, this 'governing body is a governing body is a governing body'.

## Find the images that work

The body image is an important one for us all as governors and heads to hang on to, not least as a powerful reminder of our need to work together, corporately. We noted in Chapter 3 the somewhat mechanical imagery of the government, and that, for some parents, governors and heads, life in education seems constricted, and that power and authority are experienced as measures of aggression and control rather than as resources for enabling and educating. We noted

time for something to become part of the fabric,

and

now the DfE has the stranglehold over schools.

Each governing body needs to find and use the insights of imagery. Recognize the limitations, but aim to find the image or images that work for you, the image of the school, the image of the governing body. The governing body's imagery must match the imagery of the school's learning and life. We have hinted at the dangerous reality behind the images of puppet and dummy. We recommend two images that connect to quality and accountability:

**to navigate,**

**to carry the can.**

In the chapter on accountability we offered the image of a net. Some people talk instead of, 'a machine, getting into gear, engaging the wheels and cogs', and so on. The weakness of this as an image for governing, is that the energy comes from one source, one wheel turns another, there is a single, planned interlocked system. The whole point about the net is that it is flexible: each of the people (who *are* the different strands) has some authority, energy and drive, in their own right.

The shape of the net at any one time depends on how the balance of pulls has worked out. Change is possible, not because the driver of the machine says so, but because different combinations of pull are possible on different

117

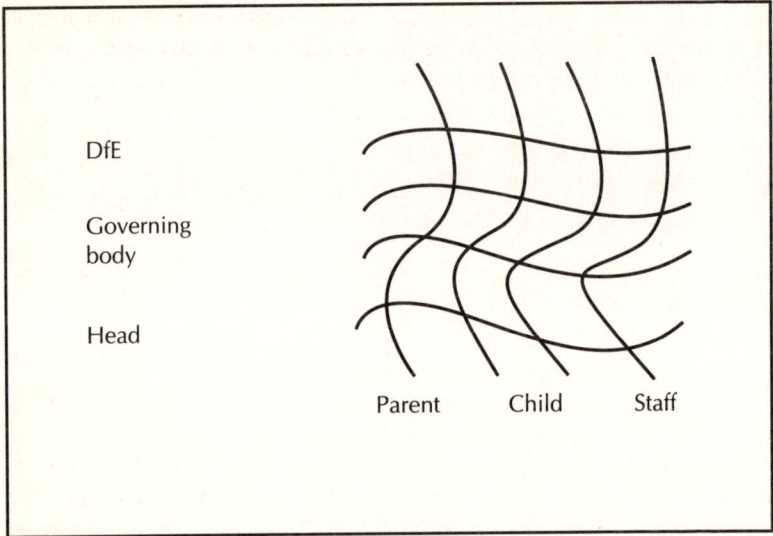

DfE

Governing
body

Head

Parent          Child          Staff

**Figure 6.2** *The net of accountability (b)*

issues at different times. You may notice how, in comparison with Figure 5.1, 'The net of accountability (a)', we've changed the place and names of some participants in the net shown in Figure 6.2. What matters is that you as head or governor can find an appropriate way to use the image of the net for seeing the connections.

## Boundaries, not barriers

In this section we bring together our own thoughts about ownership, from Chapter 4, with the government's themes of increasing parental choice and of greater accountability.

### On the boundary

We suggest you think of the school with a kind of psychological-cum-legal boundary zone around it. Just the fact that it is a particular school creates that outer edge of identity. The government's delegation of more decision making to the governing body, by-passing the former role of the LEA, has made much more distinct that sense of identity. The government's other measures to widen the information and the options available to parents, again strengthen the distinctive identity of each school. It's no longer the LEA that allocates a place in one of its schools, but a particular school which chooses to accept (or not) each parent's preferred selection of a place.

We have argued that the governing body, in facing the question, 'Whose is the school?' needs to reach its own definition of the school. We showed that *our* definition of the school would include within the boundary, as necessary parts of the identity, the four bodies of pupils, parents, head-and-staff and governors-and-head.

We have argued also that the governing body is both representative, in itself, and representing, in its activity. It is *representative* in its membership, of parents, teachers, the local community, but not yet of government. In bringing points of view into the governing body (eg, from the parish council about the use of playing fields) and in reporting from the governing body (eg, in telling local employers about new courses in technology), the governing body *represents*. For us the representativeness and the representing across the boundary gives authority to the governing body. That authority justifies the governing body in taking ownership of itself, and in taking ownership, responsibility and accountability for the tasks of navigating for the school and carrying the can for the school. Part of the accountability is to the parent body, which is *within* the identity of the school; part of the accountability is to the LEA and to the government, which are *outside* the boundary. The need to exercise the governing body's own intrinsic authority, together with the authority delegated by the government, helps us to see that *the governing body's place is on the boundary*.

Since the governing body has to navigate, its authority and time must not get immersed within the action and detail of school life and learning; the head and the school's internal management team must manage these. In its authority in navigating and in carrying the can, the governing body must be part of the school; and the proper point is where the school's identity meets the outside world, ie, on the boundary. You may notice that in Figures 2.1 and 4.1 (pages 49 and 75), we have put the governing body *on* the boundary.

This is where the idea of boundary is so much more powerful, more creative and more useful a frame of reference than the idea of barrier. Barriers get pictured as a fence, wall or gate. Barriers raise problems of access, of maintenance, of defence, and of control. 'Boundary' is a helpful notion: it causes passionate feelings of commitment and ownership, yet it is crossable and negotiable. Sometimes representatives of the governing body engage inside the boundary, in the school's action, eg, when making a staff appointment. Sometimes staff are active outside the boundary, eg, on work-related curriculum. The parent body, as part of the school, is inside the boundary. The individual parents, as choosers of the school, are outside; sometimes individual parents come inside, to learn, to be informed, to give time. The head's authority is *in* the school. It is also *on* the boundary with the governing body, and can be active *outside* the boundary, eg, as a member of the heads' professional association, or lobbying the government or the LEA on the funding formula.

## Exploring the idea of boundary

As you explore for yourself the idea of boundary, give some thought to:

- the links and connections that reach across the boundary of the school;
- the boundaries *inside* the school, of smaller areas of authority, eg, the staff body, the individual teacher's classroom;
- how boundaries expand (perhaps as the school provides a dining room and library for the community) and contract (perhaps when the school has to give up the residential class visits to the Forest of Dean);
- the importance, particularly when boundaries have to be crossed, of giving advance notice, of allowing time, of negotiating explicitly.

It may be that as heads and governors we need to educate the government about the particular realities of governing schools. We need to help government to understand and respect the notion of boundary, a notion entirely compatible with the imperatives of parental choice and greater accountability.

## Beyond the Boundary

That connection with government underlines an earlier point we made about the governing body's third way of taking ownership, namely to think wider than the school. On the boundary, you can see and you can deal both ways.

- Find your authority.
- Develop your powers and partnership.
- Think wider than the school.

Each governing body has a stake in the whole world of school government.

We highlighted in Chapter 4 the need for that world to create a mechanism for negotiating with government. We suggested that in order to mature beyond a dependence on government into an interdependent partnership with government, the 25,000 governing bodies need an identity. They need a boundary. They need a link as a mechanism to government and to other bodies, for example to teachers' professional associations. That link at present is missing. No body exists that is democratically representative of governing bodies, and engaged in representing across the boundary of school government; Figure 6.3 illustrates the point of a missing link. You may notice that in this second version of the missing link we have drawn the boundary and we have put the missing link *on* the boundary.

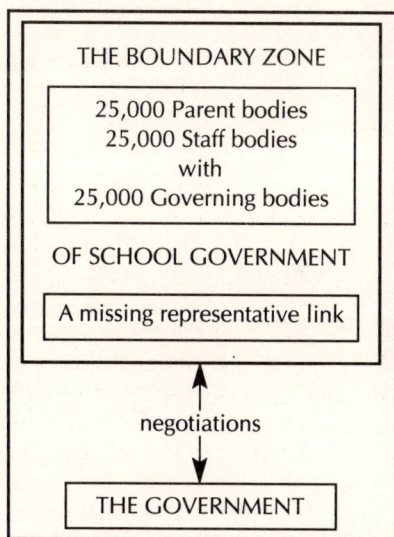

**Figure 6.3** *A missing representative link on the boundary*

The missing link needs to do for the world of governing bodies something of what each governing body does for its school.

## Being related, making a relationship

In this section we connect our comments from Chapter 2 about political behaviour, with the working links and relationships in and around the governing body.

We tried to identify in Chapter 2 the energy that is fuelling the government's moves, energies and driving forces at school level, and the driving forces elsewhere in the education system. We offered a definition of politics as the negotiating and expressing of a set of values, as the exercise of responsibility. We suggested that in schools and governing bodies there is not enough political behaviour, in that the arguments about the values of education are too few and too limited.

Now we're offering what seems to us a crucial distinction. We think it can provide a useful frame of understanding for heads and governors, and thus reduce many real confusions and unnecessary conflicts:

**RELATEDNESS**

**is**

**different**

**from**

**RELATIONSHIPS**

'Being related' is the permanent connection in the way things are. In the case of the child and its parents, the head and the chair, the teacher and the classroom assistant. There are in each instance two parties existing together, with a situation in common. 'Relationship' is what each pair makes, at any one time, of their connectedness, the fact that they are related. Relationship is about the temperature and the particular characteristics and quality of their life together, today. The child may be rebelling this week, the parents may be cold-shouldering next week, so the relationship varies, but the relatedness, the bond, the common existence, remains. This head may always defer to the chair, the chair may treat the head as a drinking companion; but a different person in the post of head, a different person in the seat of chair will make a different relationship out of the same situation that finds head and chair related, like it or not.

Relationships are not the only things that change. There used to be three related units, DES, LEAs, and schools:

DES

100 LEAs

25,000 LEA schools

The government is changing that relatedness:

DFE

1,000 GM schools      and      24,000 schools under LMS in LEA areas

The reason we recommend you to use the distinction between 'being related' and 'making a relationship', is that *each one requires a different way of thinking*. We spell this out in Table 6.2.

**Table 6.2** *Different ways of thinking of relatedness and relationship*

| |
| --- |
| You need to *see* how you are RELATED,<br><br>and<br><br>you can *decide* to draw *new connections*<br><br>but |
| Your *behaviour* makes the RELATIONSHIP, |
| and<br><br>'their' behaviour matters vitally, of course. |

## Finding the relatedness first

An unavoidable relatedness is that of the governing body and its clerk; the governing body can choose to relate to an adviser, but it cannot *not* have a clerk. If the chair and head simply tell the clerk what to put on the agenda, that's one kind of relationship; alternatively, if the chair and head develop the clerk's contribution as a valued responsible partner, that behaviour makes for a different relationship. In either case the role and task relatedness came first. You as a governor need to note and map the people and groups to whom you are related (like it or not); your map might look like Figure 6.4.

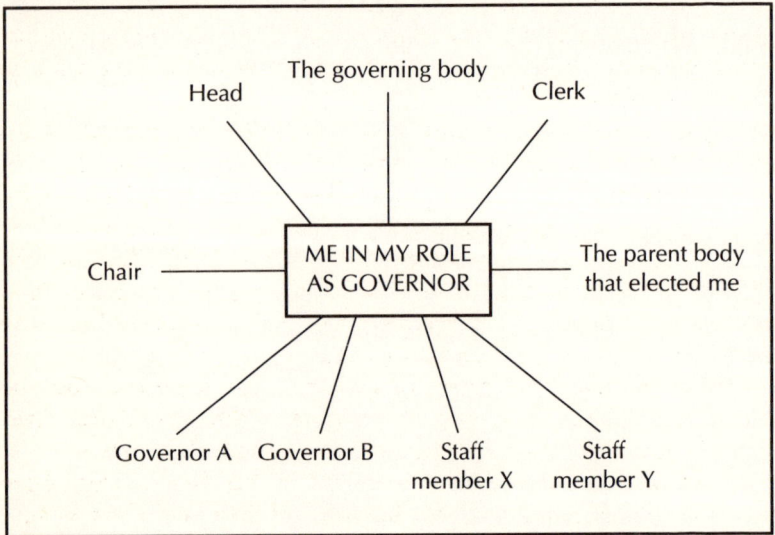

**Figure 6.4** *Map of a particular governor's relatedness*

It's up to you to decide as a governor whether or not you want to add a new link, draw a new working connection, for example to the LEA governor trainers, or to the chair of the friends' association. Like you, the governing body needs to see, map and know its connections and the groups it is related to (like it or not):

● the bodies outside;
● the staff as individuals;
● the staff as a body;
● the parents as a mass of individuals;
● the others represented on the governing body;
● other governing bodies;
● the LEA.

The governing body needs to perceive its relatedness to the Government. The governing body needs then to decide whether, at this time, that relatedness centres on obedience. '*We should* do this, because the government says so, with its authority, and it expects us to obey. Shall we submit and obey?' Or does the governing body's relatedness to the government centre on the idea of an interdependent partnership? '*We choose* to do this, because we have some authority of our own. Shall we agree or shall we disagree?' In reality many governing bodies have chosen to disagree, not over a particular principle but over timing.

For you, operating as an individual governor, the context is the governing body located on the boundary zone of the school. It is the governing body with its identity located in child-centred tasks that provides the frame of reference for you, for example, in dealing with the head and a particular member of staff over a complaint which a parent has made; for your dealings with the clerk and chair on items to go on the agenda; for the dealings between you, another governor and the staff body in a planned visit and discussion to look at equal opportunity issues.

Without the frame of reference of a common task to connect you, you will get lost in a mass of dealings that are simply based on personal lines, without common ground or common sense, with you as the centre.

You as the centre of the picture would not in itself be a problem if you had all the authority on your own, but as a governor you're not on your own. You share authority with every other individual governor. Each is at the centre of her or his own picture; that's the problem. The common frame of reference has to come from working in the governing body and in sharing its identity, and its task for the children. It is *the same frame of reference for everyone*. That's what lessens the confusion and the difficulties and makes it possible, for instance, to insist that the head leaves the meeting, with a proper 'interest', when his or her salary is being reviewed. That's what enables the governing body *not* to be preoccupied primarily with keeping the head happy and enables you to challenge the chair publicly for not keeping the governing body's eyes focused on the particular necessary decision, regardless of whether, at a personal level or on other business items, you and the chair have been 'friends' or 'enemies'. Seeing and working in terms of relatedness to a common task makes it more straightforward for the head and governors to:

● accept differences
● see the harmlessness and importance of conflict (if it's on values and on plans)

- negotiate
- need to meet.

Turning the map of relatedness into the appropriate living tissue of relationships needs behaviour that:

- thinks and plans sensitively
- gives time and fixes occasion to meet
- makes real contact in meetings.

## Relationships need to be made

Your governing body should consider the basic style of its key relationships. First you will need to *see* to whom and how you are related and you will need to decide when new connections need to be drawn. Then it will be your behaviour that makes the relationships. Will it be open and accepting, or closed and concealing? It takes time and discussion to make the connections and develop the behaviours.

## Three major relationships to make

One major relationship is that with the head; it needs much time.

A second is based on meaningful consultation. The Inner London Education Authority did it this way when reorganizing a group of schools:

> There were six stages in the consultation process. Stage 1 was a bald statement of the facts. Stage 2 consisted of a document setting out the whole range of possibilities, including the possibility of doing nothing. Stage 3 was a discussion, carried out by officials, with staff, governors, parents and anyone else who had evenings free. All concerned were then invited to send in their views. Stage 4 consisted of draft recommendations by officials, circulated to everybody who had been previously consulted and to the councillors who would have to decide. Those councillors would now engage with any members of the public who wanted to talk to them about the different alternatives. Stage 5 consisted of recommendations to the committee with all the written comments received. Stage 6 consisted of the appropriate committee making up its mind.

What are the appropriate equivalent consultation arrangements for the heads and governing body at work?

In a third area, there's 'How do we seem to relate to the outside world?' The Director-General of the CBI had this view:

> What concerns us is that almost five years after the Education Reform Act we continue to hear sounds of battle from the front. Every day comes news of a new skirmish. For many people outside the education system, the constant disputes between local education authorities, teachers, their unions, the Department for Education and Science and the various acronymical creatures of the Secretary of State are baffling at best. They create the impression that the educational establishment is more interested in its own internal disputes than it is in the fate of the children in its care, or the relevance of its activities to the prosperity and well-being of the nation. We yearn for a less confrontational approach.

How does the school seem to its outside world?

## Authority and power

In this section we link our suggestions in Chapter 5 and the government's key theme of greater accountability with the distinction we want to make between authority and power. We suggest you think of:

authority as legitimacy or right

and

power as capacity or energy.

The governing body has responsibilities. Some are for action, some are to be the line between, to be accountable. The governing body, we have suggested, is 'the vehicle for accountability' for all that the school is. It is responsible for giving account, and is liable to be held to account; it is also responsible for receiving an account from others, and for holding those others to account. We have suggested that the majority of governing bodies' lines of account relate it to three bodies: the Secretary of State on behalf of the government and the nation; the parent body on behalf of the children; and to the LEA on behalf of the local community. We see three sources of authority which justify the governing body carrying out of its responsibilities, and legitimate it:

- the government – specific authority delegated through Acts and regulations

- the local community – non-specific authority which comes through being representative
- the needs of the school – because the governing body's reason for existing is the better education of the children.

## Authority is legitimacy or right

The question, 'Is the governing body exceeding its authority?' can be answered by considering any particular problem in the light of the three sources of authority. In the cases where the law was the source of the authority, the law courts are available as a mechanism for testing legitimacy. One governing body used the High Court to pre-empt an LEA's plan to move away a temporary classroom, at least until the LEA had consulted in the manner to which it had committed itself in writing. An Industrial Tribunal, in hearing an allegation of unfair dismissal from a member of staff, does not represent a failure for the governing body; the tribunal's purpose is to see if a balance was held between the governing body's motives and procedures, and the legal rights of the employee. 'Did the governing body go beyond its authority?' This same question is at issue in an appeal panel arbitrating on a case of admission, or on a case about provision to meet a child's special educational needs. In all three examples, *procedures* are being examined. Did the LEA or governing body, in doing what it had in principle the right to do, behave reasonably in how it went about it? If not, then its right in principle could be over-ruled by its improper conduct in a particular case.

Governing bodies and heads – and LEAs and government – are often reluctant to use procedures. Procedures are intended to protect the persons or group that are properly subject to authority. Governing bodies need to learn about existing procedures, devise new ones if necessary, practise them (practising a disciplinary hearing is as important as practising a selection interview) and publicize them to all who might ever need to know or use the procedures.

Of course, procedures may slow you down; safety usually does. Procedures may mean you don't get all that you want; fair dealing means that what others want also has to be given due consideration. Of course, you can get things done if you have more power than the other persons or group. One key to good governing, however, is only to use the power that fits your authority, not to use power that feels brutal to those on the receiving end. If it feels brutal, it is probably beyond legitimate authority. Negotiated and agreed procedures reduce the scope for an unauthorized use of power.

## Power is capacity or energy

Power is around in various forms; here are some examples:

- Position, or office, I hold.... as chair, as clerk.
- Physical tools I hold.... sitting at the head of the table, holding a microphone at the APM.
- Personal abilities and knowledge I have
  - knowing about content.... entry requirements for colleges of FE/HE;
  - knowing about process ... the timetable in seeking GM status;
  - skills ... in interviewing;
  - physical characteristics.... a firm strong voice.
- Projected, supposed or attributed to me by other people.... 'she's usually right'.
- Ingredient 'X', my aura, charisma, impact.... 'he makes me a bit nervous'.

Charles Handy in his book, *Understanding Organizations*, using a slightly different classification, shows how each works in different ways, in its method of influence, in the psychological relationship it creates, and in the behaviour it calls for. The book is well worth getting hold of for the practical help it offers when applied to governing.

We want to high-light the concept of *the levers of power*. Levers give leverage, enable you to get things done, channel or develop your capacity, harness or focus your energy. If you have authority and *no* power, you are entitled and responsible for doing things; in practice, with no power, you can do nothing. Governing bodies have both authority and power.

We suggest that you will find it illuminating to think for a moment about the levers of power which are accessible to the head as head, and to the governing body. The levers seem to us to be accessible in principle, whether or not they are used that way in the current practice of your governing body.

| | Head | Both parties | Governors |
|---|---|---|---|
| *Position or office/function* | | | |
| being the chair | ✗ | | ✓ |
| drafting the minutes | | ✓ | |
| approving the draft | | | |
| unsigned minutes for interim use | | ✓ | |
| formally approving the minutes | ✗ | | ✓ |
| holding to account by appraisal | ✗ | | ✓ |
| holding to account otherwise | ✗ | | ✓ |
| setting targets | | ✓ | |

|  | Head | Both parties | Governors |
|---|---|---|---|
| chair's action |  | ✓ |  |
| appointment of staff |  | ✓ |  |
| managing staff | ✓ |  | ✗ |
| managing the time of the Clerk | ? |  |  |
| creation of the meeting's agenda |  | ✓ |  |
| receiving school information. | ✓ |  |  |
| *Physical tools held* |  |  |  |
| top of the table |  | ✓ |  |
| arrangement of chairs |  |  |  |
| food and drink as refreshment, reward or hospitality |  | ✓ |  |
| power of signature | ✓ |  |  |
| production of information | ✓ |  |  |
| time in school. |  |  |  |
| *Personal abilities and knowledge* |  |  |  |
| grasp of role of head | ✓ |  |  |
| grasp of role of governors |  | ✓ |  |
| knowledge of relevant law |  | ✓ |  |
| knowledge of relevant procedure |  | ✓ |  |
| knowledge of education | ✓ |  |  |
| contacts with LEA officers | ✓ |  |  |
| contacts with LEA members |  |  | ✓ |
| contact with the chair. | ✓ |  |  |
| *Projected, supposed, attributed* |  |  |  |
| traditional role of head |  |  |  |
| – as director | ✓ |  |  |
| – as expert | ✓ |  |  |
| – as .... |  |  |  |

*Ingredient 'X'*

Think about the distinction between the head as head, and the head as part of the governing body. Remember that there is one head and probably a dozen or more governors. Note that some levers are not accessible, usually by Act or regulation. Note the central column, of levers accessible to both parties, where you need to think what applies in your own context.

We suggest you will find it illuminating also to use the levers of power for thinking about the relatedness and the working partnership between the *governing body* and the *Secretary of State*. The levers you develop and use will turn that relatedness into a particular constructive or sterile relationship.

| | The governing body | Both parties | The Secretary of State |
|---|:---:|:---:|:---:|
| *Position or office* | | | |
| producing paper | | | ✓ |
| appointment of governors | ✓ | | |
| suspension of governing body | ✗ | | ✓ |
| resignation of governors | ✓ | | ✗ |
| making legislation and regulations | ✗ | | ✓ |
| funding | | | ✓ |
| publicity | | ✓ | |
| requiring reports about the school | | ✓ | |
| inspecting the school | | ✓ | |
| access to national bodies | | ✓ | |
| creating a body to represent governing bodies. | | ✓ | |
| *Physical tools held* | | | |
| creating the agenda | | ✓ | |
| voting | ✓ | | ✗ |
| time for the school. | ✓ | | |
| *Personal powers* | | | |
| experience in governing | ✓ | | |
| expertise in governing | ✓ | | |
| knowledge of the school | ✓ | | |
| access to the MP | | ✓ | |
| actually transacting business in the governing body. | ✓ | | ✗ |
| *Projected, supposed, attributed* | | | |
| respect for the rightness of 'the law' | | | ✓ |
| goodwill and confidence from the parents. | | ✓ | |
| *Ingredient 'X'* | | | |
| *The media* | | ✓ | |

It is clear that the last few years have been characterized by the Secretary of State's acquisition of powers; but, more significantly for us, these last few years have also been characterized by the inability of governing bodies even to think about the levers of power they might develop.

## Developing powers, capacity and energy

Perhaps your governing body could consider the following:

- Why do people dislike *the word 'power'*?
- Do some people *want to be powerless*?
- *What disempowers* us?
- *What training and practice* will help heads and governors to use power within the scope of the governing body's authority?
- *Is the governing body meddling* in the head's day-to-day management of the school?
- Does the head *patronize* the governors?

Remember that in the authority for school governing, little is solo, most is shared.


# Dimensions of the governing body's contribution

The previous section ended with the warning that, in the authority for school governing, little is solo, much is shared. This section offers a framework to enable your governing body to take an overview and to bring balance to its unique contribution.

We have already offered a working definition of partnership as a checklist for the head and staff, governors, the parent body and the LEA, to use as common ground for their negotiating on particular needs and tasks:

> a working relationship that is characterised by a shared sense of purpose, mutual respect, and the willingness to negotiate. This implies a sharing of information, responsibility, skills, decision-making and accountability.

Another set of questions for checking if you are working on the lines that matter is to ask whether the governing body's contribution is being accounted for, engages it with others, is practicable and is distinctive.

We suggest that there are six dimensions which seem necessary to take in all that the school is, and to meet the governing body's responsibility and accountability. We list them below, with examples that were uppermost in governors' minds at conferences in the spring and summer of 1993.

- **Advising**, eg, in telling the government about testing at Key Stage 3; in telling the LEA about the linking of pay policy and staff performance.

- **Policy-making and development-planning**, eg, in considering GM status; in balancing the budget.
- **Coordinating and mediating**, eg, seeing how our personnel policies compare with other schools and the LEA's guidelines; in bringing together funds for the school's environmental centre.
- **Supporting and promoting**, eg, in giving staff sufficient time for the necessary paperwork; in letting the local community know what the development plan priorities are.
- **Monitoring**, eg, how our policy for special needs is working in practice; how pupil numbers on roll are matching our forecasts and the plans based on them.
- **Governing body's working methods,** eg, is poor chairing the reason why early items crowd out the rest of the agenda; what can we do about 'the old hands not listening' to the new governors?

For each of the six dimensions the most important area of work is that your governing body should decide for itself. What we want to emphasize is the width, the overview, the taking in of all six dimensions. You will find these ideas worked into a pack, *Self-Appraisal for the Governing Body*, produced jointly by NAGM and AGIT.

## A 'whether' forecast

Most weather forecasts have a general situation report, a forecast of probabilities and some mention of possibilities. In this section we summarize the present situation, note some trends and suggest probabilities. Each probability depends on 'whether x or y.'

### The general situation yesterday: responsibilities and influences

Central government is busy with laws, regulations, inspection and advice. Heads are managing the implementation of policies, the day-to-day running of schools, school-based reviews and leading their staff. Governing bodies are directing the aims of the schools and the control of finance, discussing policy on the whole curriculum and some specific curriculum areas and managing the appointing of staff. LEAs are setting local curriculum aims, offering guidance and administrative seminars and framing budget levels and staff employment policies over all.

## The developing trends

- The LEA is losing functions.
- Central government is controlling money and revising a common funding formula.
- LMS requires more energy.
- LMS has been introduced with no additional resources. Governing bodies are likely to have things to say about resourcing.

## Probabilities and possibilities

- Government may be accepted as the only operating source of authority.
- Governing bodies may establish a credible national voice.
- Government may be no more responsive than LEAs were.
- The system, the arrangements and the requirements, cannot be policed nationally on a regular basis.
- The political drive, since the early 1980s, to make educational policies may be followed by a different thrust.
- The system will become looser-jointed as choice and diversity increase.
- There is scope for local energy and initiative beyond obedience to bureaucratic directions.
- There is scope for local negotiation and for amalgamation or federation, for different purposes and at different levels.

## It depends whether ...

- Governing bodies develop and act on their authority as representatives and in the face of school needs – rather than simply colluding with authoritarian instincts.
- The government (eg, in the School Curriculum and Assessment Authority and in the Funding Agency for Schools) engages with governing bodies as fellow-responsible-managers – rather than treating the latter as inferior cogs or agents in the centrally-driven mechanism.
- The diversity makes the government so uncomfortable that it tries to police by even more regulation, failing to see that proper policing needs people and that regulation is paper.
- The government (or the nation, or the European Union) makes space for independent local government.
- Governing bodies learn to negotiate with the LEA and each other and government, and whether heads and governors learn to negotiate between themselves.

● Governing bodies are willing to interpret the law, rather than ask the government for more definitions, and settle for low-level administrative management.
● Governing bodies can be creative in meeting the needs, locally and nationally, rather than painting by numbers the scenes they're given to paint.

Another way of putting it is: **whether the government listens to what governors say.**

## We say

If we as authors could speak on behalf of heads and governors, our messages to the government would be:

● Don't make us so sick of change; give us a respite, time to digest.
● Don't centralize so much that local managers become little more than technicians, oiling and operating the machine.
● Don't try policing. Give us time and space to find *our* ways to meet *your* purposes.
● Do think about partnership as we have defined it.
● Don't cage us in
● Let the system and the service grow, and cut some of those regulations.
● Help us organize ourselves to meet you on level terms for our common business. We have views too, on what makes for the better education of the children.

# References

111. 'On me as a dustbin', Helen Goff, *Cadbury Seventh Book of Children's Poetry*, Arrow, 1989
113. Robert Fulghum, in *The Early Learning Project*, Royal Society of Arts, 1992
126. 'Six stages in the consultation process', Sir Peter Newsam, *TES*, July 1993
127. 'What concerns us', Howard Davies, *The Independent*, January 1993
129. Charles Handy, *Understanding Organisations*, Penguin, 1976
133. *Self Appraisal for the Governing Body*, 1992, NAGM-AGIT, Lyng Hall, Coventry CV2 3JS

# Chapter 7

# Getting the Act Together

---

> *In this chapter ...*
> How do 'I' get to be part of 'we'?
> How do we build relationships in the governing body?
> What work do we focus on as the governing body?
> Who do we build relationships with beyond the governing body?

In this chapter we want to spell out the straightforward advice that you may already have drawn for yourself from what we've said in the earlier chapters. It's about using the picture you've made in your mind. It's about moving on from being a head or a governor simply related to other governors. It's about building relationships to create a governing body.

Getting the act together as a team matters because it creates a new whole that is more substantial than a loose collection of individuals. It increases the effective exercise of responsibility. It calls for commitment and growth in those involved, because, in concentrating on the job, they want not only to get the job done but to build the team. It's a more appropriate way in our democracy to carry accountability. We suggest two key words which help in the team work of governing:

**Boundary**

**Negotiate**

Our advice begins with you personally.

# How do 'I' get to be part of 'we'?

Think of your warmest, strongest, most valuable relationship. That's you at your best. Try to replicate that most valuable relationship in your connectedness with each other and the head. Be your best and fullest self by giving and sharing. A microcosm contains in it everything that is in the larger whole; what's in you is almost certainly in other people too: your pride, your embarrassment, your wish to improve things. Find out about the other governors and the head and what makes them tick. You may only get superficial answers to begin with. The more you share of yourself – and the more you listen, the more you'll get. Take a chance. Act on your own authority and initiative if you have to. If something goes wrong, it's usually retrievable. Some of the best managers say, 'It is better to ask forgiveness than to seek permission'. But remember that some of your behaviour may block while some of it can build.

When you disagree bluntly, when you shut others out from making their contribution, when you feel you have to defend yourself and do it by attacking others, your behaviour has the effect of putting blocks in the way of progress.

But when you make proposals or add to the ideas of others, when you widen the scope of who can contribute and what may be relevant, and when you check that everyone in the discussion is reaching the same understanding, then your behaviour builds strength for going forwards.

## Think it through

When you have something that needs saying in a meeting, think it through first. This acronym, DESC, may help you to remember what you want to say:

D   *Describe* the facts.
E   *Express* your feelings; that eases it for others, they have to use *your* words for *your* feelings; it helps you control the feelings.
S   *State* what you want.
C   *Calculate* the 'what ifs'; the anticipation makes the discussion less dark and uncertain for you. The 'what ifs' can't *all* happen; it makes you ready to negotiate.

## Getting from 'No' to 'Yes'

Don't threaten. Be prepared to persuade. Think about how to get from 'No' to 'Yes'. Table 7.1 draws suggestions from a training film and booklet by Video Arts.

**Table 7.1** *Getting from 'No' to 'Yes'*

| |
|---|
| *Win yourself a hearing*<br>–     Explain your own feelings.<br>–     Refer back to the points of others.<br>–     Make your points firmly, but stay friendly. |
| *Listen actively*<br>–     Show them you understand ...<br>      – that they feel strongly,<br>      – what they feel strongly about,<br>      – and why they feel strongly about it.<br>–     Pause to let them respond. |
| *Work to a joint solution*<br>–     Seek their ideas.<br>–     Build on their ideas.<br>–     Offer your ideas.<br>–     Construct the solution from everyone's needs. |

Remember that 75 per cent of suggestions get accepted, in one way or another, in due course. It may not be exactly your way, nor your time-scale, but in the governing body you are committed to working to joint solutions. If you are isolated on an issue, it's clearly not the climate or not the time for this governing body to move in the direction you are suggesting. Things can ripen; you've planted a seed, tend it quietly for later.

When you have something that needs doing, think whether it is the thing you most want to give time to and whether it is practicable. Discuss it with one or two other governors; talk about it with the head; let the head and chair know you've asked the clerk to put the item on the agenda. Ask for the governing body to set up a working group with authority to consider and to recommend. Don't ask for authority to do or to decide, which may seem threatening and disruptive for the governing body. Know who you'll recommend to be on the working group. Don't respond to the comments in the initial discussion; just note all the points and promise to consider them. Be prepared for confrontation of real issues and feelings. Education is about values that matter and all of us as heads and governors feel strongly. Play your part in keeping the conflict focused on the ideas and details, using different ways to resolve the conflict. Below is advice from a book with the encouraging title *Everyone Can Win*. It lists, in no particular order, a range of suggestions about what you could say, how you could change your behaviour and how you might respond when confronted.

> Ask a question
> Let some hostile remarks pass
> Rephrase hostile remarks
> Write down what is being said
> Ask others to rephrase, positively
> Uncover the tactic by telling them you can see it
> Speak clearly
> Agree to discuss it later
> Use humour
> Change tack
> Call 'time out' (suggest a break in the meeting)
> 'Show me how that's fair'
> 'Please tell me what you heard me say.'
> Can it be dealt with privately?

What's unhelpful is *not to admit* the clash, and *not to resolve it*. Mapping people's positions and arguments is a help. Table 7.2 offers a simple format for noting the key points, what the issue is, what's involved, what each needs and each fears. You may like to use it as private notes, or offer it to the governing body to help clarify an important or awkward argument. Putting it on a flipchart may help.

Getting their arguments out into the open is a help to most people. They won't feel squashed personally. They know everyone can see and think about their arguments. People often don't mind losing arguments, but they mind, very much, not being heard.

## The head and the chair

Prue Denton and Eric Benton, chair and head of a primary school, recommend the following points for any chair and head to consider with the aim of improving the quality and effectiveness of their relatedness:

Can you –
●        accept one another's strengths and weaknesses?
●        be discreet and not break confidences?
●        be honest with one another?
●        be loyal to each other and to the school?
●        respect one another?

**Table 7.2** *Mapping the conflict*

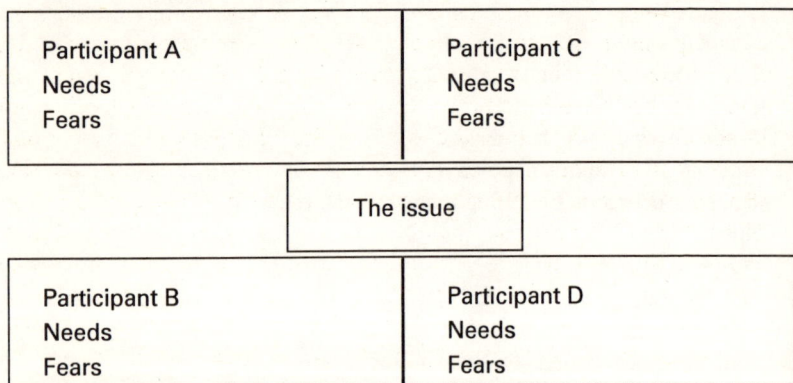

| Participant A<br>Needs<br>Fears | Participant C<br>Needs<br>Fears |
|---|---|
| The issue | |
| Participant B<br>Needs<br>Fears | Participant D<br>Needs<br>Fears |

- share the responsibility and the workload?
- trust one another?

Any governor, except the head and any one employed by the school, can be chair. Perhaps everyone *ought* to take a turn, including you! Members can be trained to be chairpersons. Former chairs are useful for chairing committees. There is an important sense in which every governor needs to think about and feel in touch with the leadership and executive roles of head and of chair.

## How do we build relationships in the governing body?

As a governor or a head you are in politics, negotiating and expressing a set of values and exercising responsibility; that's what governing is. We suggested earlier a crucial distinction: relatedness is different from relationships. You have a relatedness to all the governors and the head; your behaviour makes or mars the relationships. The focus is the overall task that brought you all together, which is the better education of the children.

There seem to be four areas of relationships for us all as heads and governors to think about and to work at; these can be summarized as:

- Making common maps.
- Focusing on specific key tasks.
- Creating and developing mechanisms.
- Remembering how groups behave.

## Making common maps of values and aims

Set aside time as a governing body to identify your different values and aims; this may take half a day each year. Then make time to resolve some of the differences into an agreed common basis for the school, using common images of the school and of the governing body. Draft and refine definitions of who is the school and whose is the school. This lays the ground for all your subsequent work, provides the assumptions for any development plan and gives a yardstick for prioritizing your agendas. As you get used to these fundamental exploratory discussions, you'll get the confidence to share your thinking, with the staff body, the parent body, perhaps even the pupil body. You need to keep clear that the governing body belongs to all of you. Watch that A and B teams don't develop with a few individuals doing all the important work. You need leadership, yes, but keep it accountable to the whole governing body. Get contributions from all for the common maps and make sure that the representative governors offer some comments from their particular angles. Don't accept silence in these fundamental discussions.

## Focusing on specific key tasks

The relationships that matter are the ones that grow from working together on real issues. Remember that you are operating at the level of the overview. What we said about the levers of power, in Chapter 6, should give you the confidence to mobilize energy for these daunting key tasks, provided you decide to let the less important business wait, or delegate it. Work out the hours available each term from the governors, then allocate time and particular individuals to the priority tasks.

## Creating and developing mechanisms

Creating and developing mechanisms is about making and using levers of power. These include:

- groups
- documents
- occasions
- authorized individuals
- sessions for governing body development.

We remember reading of a chair who was resigning. His reason was simply a lack of energy and time to keep confronting the head on what felt like almost every issue. It seems to us that the individual made the understand-able mistake of seeing it as only the chair's responsibility to deal with the head. It seems that the individual as chair failed to get the governing body to mobilize and authorize the other governors, including the teacher-governors, into a network of working groups and authorized individuals dealing with separate issues. Carrying all development on the one-to-one, head-to-chair basis was impossible, hence the resignation. It would have been more appro-priate to use the head-and-chair relatedness as a steering-point, a safety valve for problems rather than as engine-power for doing all the work.

## Working groups as a mechanism

Working groups mobilize energy and get recommendations drafted more easily than can the governing body as a whole, and more corporately than an individual can. Working groups may be appropriate for single short-term tasks like drafting the self-appraisal and annual target-setting of the govern-ing body, or engaging with the staff in drafting the development plan. A standing committee may be more appropriate for more technical and more continuous policy matters, such as pay, monitoring school quality, and per-sonnel policies. Working groups need very simple terms of reference, to investigate and to recommend. They need at least one or two people with some knowledge and passion for the specific task and should be encouraged to seek advice from others. They provide an invaluable opportunity to include one or two new governors who can gain confidence by working in a small group and by producing an immediate achievement. Think who could offer a relevant professional view, who could offer a common sense per-spective. Think, too, about the different kinds of team-roles that people like to take and are good at: the coordinator; the shaper; the completer-finisher; the one with ideas or contacts. Get a wide range of skills into the working group according to its task.

## Documents as a mechanism

Documents help people to focus. Written recommendations let everyone see what the issues are. Clear and brief reports enable you to disseminate widely to influence many others. Last year's development plan provides a common visible starting-point for work this year. Draft minutes (as agreed by clerk and chair for working purposes) can let parents and staff learn from their notice boards where the governing body is in its thinking, as well as remind-ing governors who has to do what. The school brochure, on the coffee table in the waiting area, lets visitors see at once what the school's values and aims are. Information is power; documents are levers of power. Imagine the local impact of this statement from the Dorset Education Partnership:

We, the governors, parents, headteacher and staff of Elysian Fields School believe that we should work together in close partnership in accordance with the following principles and practices:

- We all share in one common purpose; to work for the benefit of all pupils and to ensure that the school is used and valued by the community it serves.
- We are committed to working openly and democratically, so that all interested parties can understand how the school is managed and how decisions are taken.
- The governors and parents value highly the professional roles of the head and staff.
- The headteacher and staff warmly welcome the full involvement of governors and parents in the school.
- The governing body will encourage the closest possible links between parents and governors and will give priority in its meetings to issues raised by parent-governors.
- Parents will keep the school informed of all important matters relating to their children.
- The headteacher and staff will likewise keep parents informed on all important matters.
- The headteacher and staff will arrange regular opportunities to meet parents and review each child's progress.
- The governing body, headteacher and staff will keep home-school liaison and community links under regular review.
- The governing body, headteacher and staff will consult parents on school policies and priorities for development.
- The governing body, headteacher and staff will seek parents' views on the quality of the work of the school.
- The partnership will work best if problems or difficulties are shared openly and speedily with whoever is immediately concerned and able to help.
- The partnership will encourage the collaboration of governors, parents, headteacher and staff in all aspects of the education and welfare of children in school, at home and in the community.

Can you set out to negotiate an equivalent for your school?

## Occasions as a mechanism

Make occasions. We've written elsewhere about the potential we see in the Annual Parents' Meeting. Why not two well-advertised one-hour sessions each term, for the governing body to talk with available staff and interested parents about one or two key agenda issues that face the governing body? Why not a termly one-hour exploratory meeting for the governing body with *all* staff, seeking views on personnel issues, and testing proposals on aspects of organizing the life of the school? The fact that such meetings happen, that they are planned and regular, that they table one or two issues that are among the governing body's current priorities, will build the relationships as well as help to get the tasks done. Why not a ten-minute slot, at the end of each governing body meeting, to review how well the governors and head worked in the meeting? Attending to behaviour is the way to get things done, *and* the way to keep the given relatedness growing into constructive relationships. At one school, two governors, a different pair each time, take appointments on parents' evenings to discuss matters appropriate to the governing body. It acts as an early warning system.

## Authorized individuals as a mechanism

The governing body can't do it all, not even with committees with delegated authority, nor even with investigative working groups. Some things need to be delegated with authority to individuals. For most governing bodies this happens, accepting the head's executive position and by using the discretion of the chair. 'The Chair ... has power to discharge as a matter of urgency, any function of the Governing Body of the school'. That's an enormous lever of power. The School Government Regulations of 1989 which grant that power also say, 'The Governing Body may resolve to delegate ... to any member of the Governing Body, or to the Headteacher if he is not such a member, (almost) any of its functions'.

Most governing bodies in our experience don't use this freedom to delegate, perhaps because they don't know they have it. The governing body needs to choose carefully the person, and define clearly the scope and the timing of the delegation. Inappropriate authority and power could distort the net of inter-meshed responsibilities and interests; particularly, the authoritative role and powers of the head can skew the balance. Head-and-chair as a pair have even more authority and power. It is crucial that authority and power are consciously delegated rather than picked up by default.

*Sessions for governing body development*

This is the last mechanism we wish to highlight. We've mentioned already the half a day a year to discuss and review values and aims, the ten-minute review at the end of each meeting, and the annual self-appraisal and target-setting of the governing body. (This target-setting is for what the governing body *itself* will do; it is just one piece within the greater jigsaw that is the development plan for the school.) Within the range of training possibilities that focus on the governing body, we mention just two more:

● Have a consultant or mentor to work on a termly basis with the head and chair as a pair. Their responsibility is to offer the major leadership needed by the governing body. An hour a term with another head or chair, or an LEA officer or an inspector, would be illuminating.

● Try this exercise once a year at the beginning of a governing body meeting. The head and governors are grouped in sixes. Each individual in turn speaks briefly on a topic. All six then write down and give back to each of the other five what she or he heard them say. It takes only a few minutes – and proves how selectively people listen.

If the governing body devotes time to its own development, it will come to do its task better. It will also give the head and governors more satisfaction and more conviction about the worthwhileness of the governing body. It is the governing body that has a low sense of self-worth that tries to spend its training money on things for the school.

## Remembering how groups behave

Head and governors seem surprised that behaviour in the governing body is 'different', presumably meaning different from what they might have expected from knowing the individuals. But the same individual's behaviour, in a pair, in family or friendship groups, and in task-oriented groups – which is what governing bodies are – will always differ.

From the mass of helpful writing and training about behaviour in groups, we offer you Tuckman's model of group life; use it as a reminder of what's normal about groups when your governing body seems to be struggling. The model describes a group progressing through five stages. Many groups progress sequentially; however, some groups may not pass through all five stages, while others may jump backwards and forwards from one stage to another:

*Forming*. The group is characterized by anxiety, with a dependence on the leader. The group will be attempting to discover its code of conduct, testing out what behaviours are acceptable. A group in this mode will approach the task with the feeling, 'What shall we do?'

*Storming*. The group feeling will be one of conflict, with rebellion against the leader, polarization of opinion, conflict between sub-groups and resistance to control. There is likely to be emotional resistance to the task and a sense of the task being impossible. A group in this mode will often approach the task with the feeling, 'It can't be done', or 'I won't do it'. This stage often represents a testing-out of the leadership. Viewed positively, this may be the means by which the group starts to take the task seriously.

*Norming*. At this stage some group cohesion develops and norms for the group emerge. Former resistance starts to be overcome and conflicts are patched up. The group in this stage will be capable of offering mutual support to members. Increasingly a determination to achieve the task will be accompanied by an open exchange of views and feelings and a sense of cooperation. A group in this mode will often approach the task with the feeling, 'We can do it'.

*Performing*. The group is doing the task which has brought it together. Roles within the group are functional and flexible, with interpersonal problems having been patched up. Individuals feel safe to express differences of opinion, and trust the group to find acceptable compromises if necessary. There is lots of energy available within the group for doing the task and, as a result, solutions to problems emerge. A group in this mode will approach the task with a feeling of 'We are doing it'.

*Ending*. Individuals leave the group, and discussion focuses either around past shared experiences or suggestions to hold the group together. Often suggestions will be made for the group to meet at a later date for further support. There is a desire for the group not to end completely.

A key permanent factor is tension; its absence signals slackness, which means that work does not get done. The governing body needs to hold the tension of people being open and issues being confronted. Work with cooperation *and* conflict; map the conflicts as we suggested earlier; face up to mistakes. Trust will grow out of errors and regrets, as well as from successes.

## Ground rules for talking

The governing body works a lot with words so you need to establish ground rules for the way you talk. A class of top infants came up with the following rules for their talking:

- Don't keep comments to yourself.
- Help each other discuss.
- Keep trying.
- Cooperate. Stick together.
- Don't leave anyone out.
- Give information.
- Don't shout.

# What work do we focus on as the governing body?

The general governing body is a local community to provide for leadership and accountability; operational executive matters should be delegated to the head. One test of the leadership and accountability is whether what's achieved improves the education of the children. Another is whether the work could not or would not have been done without the governing body. A third is in the judgement of head and all governors about their corporateness. So, thinking of community, leadership and accountability, is one way to decide what work to focus on.

You may feel that we're ducking the question, 'What do we *do* as head and governors?' The difficulty is in making sense to 25,000 different governing bodies. One common recommendation is to develop a policy-making forum.

### The policy-making forum

We offered earlier the idea of the governing body on the boundary – the organizational boundary with the world. The governing body was predominantly *on* the boundary, in small part within the school, in larger part in the world outside. The head and the school's internal management team were *in* the school for action. We're now using that idea in finding a practicable combination of the governing body, the head and the internal management team, for planning and policy-making purposes.

To make sense of Figure 7.1, note first the boundary and external position of the governing body; then note the mostly internal position of the head and the internal management team and, finally, note that the policy-making forum links them *on* the boundary.

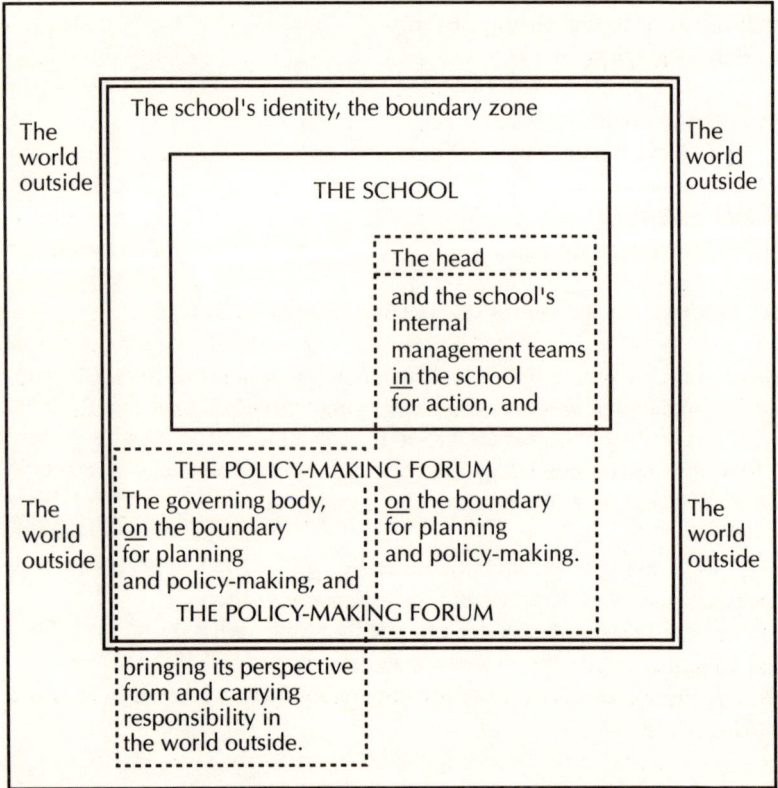

**Figure 7.1** *The policy-making forum*

The policy-making forum in no way detracts from the responsibility and accountability which are the governing body's. The forum mobilizes the expertise within the school. It reduces for the head the onus of being the only school-based planner in the governing body and frees the head to behave as a governor some of the time, confident that the executive management idea of headship is, for the time being, being carried by the deputy and other senior staff; they and the head will have had their prior discussions anyway. It eases for the teacher-governors their deference to the head as their executive manager, and allows them to use their professional attitudes and values *as teachers* to act as governors for this *particular* school.

You must decide whether the forum means bringing two or three senior staff to meetings of the governing body to represent management, or whether the forum means concentrating the governing body's meetings into two parts, 'policy planning' and 'other', so that the governing body does work on its own as well as with the school's internal management team. What we are recommending is that establishing the policy-making forum would integrate the people who have the greatest stake in planning the school's advance into an uncertain future. One way would be to build the parent body into the policy-making forum, this adding to the very valid representativeness which comes through the parent-governors; they were elected, not by the individual parents, but by the parent body. As an interim step, we know of at least one junior school which has had for some years a curriculum policy group comprising teachers, governors and parents.

## What else to focus on

In earlier chapters we pointed to certain work that needs a unique contribution from the governing body:

- Getting the values discussed, agreed, and published.
- Deciding who you mean by 'the school'.
- Thinking about *all* the dimensions of the school.
- Using the characteristics of effective schools as a checklist in the annual monitoring.
- Target-setting, and appraising the governing body's contribution.
- Working at the development cycle and plan.

## Competences

If you really mean business, we suggest you take the following list of 'competences' or characteristics required for school managers and use it in a half-day development session for the governing body. The idea of competences was refined by the National Educational Assessment Centre in assessing and preparing people for headship. We suggest that this idea of competences could be useful in the appraisal and development of the head, and in the self-appraisal and self-development of the governing body.

*Administrative competences*
- Problem analysis: ability to seek out relevant data and analyse information to determine the important elements of a problem situation; searching for information with a purpose.
- Judgement: ability to reach logical conclusions and make high quality decisions based on available information; skill in identifying educational needs and setting priorities; ability critically to evaluate written communications.
- Organizational ability: ability to plan, schedule and control the work of others; skill in using resources in an optimal fashion; ability to deal with a volume of paperwork and heavy demands on one's time.
- Decisiveness: ability to recognize when a decision is required (disregarding the quality of the decision) and to act quickly.

*Interpersonal competences*
- Leadership: ability to get others involved in solving problems; ability to recognize when a group requires direction, to interact with a group effectively and to guide them to the accomplishment of the task.
- Sensitivity: ability to perceive the needs, concerns and personal problems of others; skill in resolving conflicts; tact in dealing with persons from different backgrounds; ability to deal effectively with people concerning emotional issues; knowing what information to communicate and to whom.
- Stress tolerance: ability to perform under pressure and during opposition; the ability to think on one's feet.

*Communicative competences*
- Oral communication: ability to make clear oral presentation of facts and ideas.
- Written communication: ability to express ideas clearly in writing; to write appropriately for different audiences – students, teachers, parents, etc.

*Personal breadth competences*
- Range of interest: ability to discuss a variety of subjects – educational, political, current events, economic etc; desire to actively take part in events.
- Personal motivation: need to achieve in all activities attempted; evidence that work is important to personal satisfaction; ability to be self-evaluating.
- Educational values: possession of a well-reasoned educational philosophy; receptiveness to new ideas and change.

We hope this section has given you an intelligible agenda. Don't worry if you can't get it all going at once.

## Who do we build relationships with beyond the governing body?

We hope you will have accepted our ideas about relatedness and relationships. We recommended that the governing body map the connections that relate it, like it or not, to others concerned for the better education of children. All that we have said about behaviour and about mechanisms applies to this section, too. We want here to highlight some reminders from earlier chapters about the governing body and its connectedness, and make a few additional points.

## Within the school

The individuals and groups within the school should be the first concern; don't equate the governing body with the school. Angela Thody, in her book, *Moving to Management*, gives a convenient summary.

Our view of the school includes the parents and the parent body. One relationship to build, then, is a responsible partnership with the parent body in the APM and other policy-issue forums.

### The world of business

We quoted earlier the Director-General of the CBI's comments about the warfare that seemed endemic in education. He also made this remarkable declaration of partnership:

> Heavy and effectively targeted investment in education and training (together with enhancement of the transport and communications infrastructure that links citizens to each other and to world markets) are the CBI's top priorities in its representations on public expenditure.... *Business would like to be able to support teachers and educationalists in the public expenditure battles ahead.* As the CBI has repeatedly said, education and training are among our highest priorities. Surveys regularly show that the poor quality of our people skills lies at the heart of Britain's economic problems. But our support would be more enthusiastic if we were sure that the best use was being made of the £30bn now spent, and

that the top priority of those working within the system was, indeed, to make it work. [The emphasis is ours.]

**What government could withstand the combined power of pressure from the CBI, parent bodies, governing bodies and staff associations?**

## The local education discussion

At the local level, one mechanism is to widen the APM by inviting representatives and interested groups from the local community. Another is to tap that declared commitment of the CBI in its local forum.

We recommended giving 10 per cent of your time as a governor to work beyond the school's physical boundary. Work with the idea of the local net, and make networks; build up the LEA-area-based association of local governing bodies. While question marks may hang over the value of local government as we now have it, let's note the comment of Professor Kogan from Brunel University that, 'Only in the UK, France and New Zealand does the local authority not remain as the principal agent of public education policy'. If the government does not foster subsidiarity, that 'New European' word, then we as heads and governors need to work to build 'the new focus of democracy' (Angela Thody's words again). An interesting article by Chris Waterman of the Association of London Authorities reported from his crystal ball on what might be the effect of John Patten's Education Act of 1993 in ten years' time:

It's difficult to put your finger on what's missing, until you realize that the difficulty, in fact, is finding something to put your finger on. The major gap is where the education discussion used to be. The local authority is no longer the focus for debate about the shape or nature of the local education system. The community consensus has been replaced by individual schools negotiating individual deals with the Funding Agency.

We, the authors, believe that no governing body, no school, can do without 'the education discussion'. Governors and heads probably have to *make* forums for it: the school's policy-making forum, the LEA-area-based association of local governing bodies, and the national missing link that would connect to government, which we see as:

- a national body
- democratically representative of governing bodies
- with governing bodies as the prime membership

● with the principal task of championing the causes of governing bodies in their dealings with government.

For its accountability and negotiation with governing bodies the government has no annual report, no annual governors' meeting, no consultative official forum with governing bodies (nor with parents), no acceptance of a general teaching council, and no budget according to the costed requirements of the National Curriculum. At least on the consultative forum we as heads and governors can set the pace. We need a crucial disposition in all governing bodies and in all the present organizations to bring about a transformation, or a new creation. That, at the national level, is the macro model of what each governing body needs to do at its own micro level of governing the school.

## References

139. *Getting from 'No' to 'Yes'*, Video Arts, London, 1988
138. *Everyone Can Win*, by Helena Cornelius and Shoshana Faire, Schuster, 1989
139. Prue Denton and Eric Benton in *Managing Schools Today*, September 1992
143. The Dorset Education Partnership, David Rees, *Home and School*, NCPTA, Summer 1993
145. Tuckman's model of group life, in *Working Together*, Tacade
147. Ground rules for talking, in *A Framework for the Primary Curriculum*, National Curriculum Council, 1989
150. Competences, Derek Esp, *Competences for School Managers*, Kogan Page, 1993
151. 'Heavy and effectively targeted investment', Howard Davies (reference 127, above)
153. Chris Waterman, *Education*, 1992

# Chapter 8
# Watch this Space

This book has been about the scope for exercising responsible and account-able authority by the micro and macro political behaviour of governing bodies, creating a local directing community with the local manager (the head). Two key questions remain to be answered in the years ahead:

**Can the government let governing bodies breathe independently?**

**Can heads help governors to work on the boundary?**

The capacity for breathing independently is evident in the slow response to the government's pace on GM status; in standing with others against pre-mature arrangements for testing children; and in not readily taking up the discretion to award performance-related incentive allowances to teachers. Issues that could lie ahead include: protecting the schools against schemes for performance-related pay devised on inadequate and not widely agreed criteria of merit; protecting the children by resisting the introduction of new school-based teacher training unless we are satisfied of its adequacy; reach-ing agreement with government about the level of resources; and reaching agreement with the Funding Agency for Schools about appropriate princi-ples for distributing those funds.

The role of governing bodies can be seen clearly. The images of guardians and of trustees, responsible and accountable today and for tomor-row, are a good basis. The accountability is for quality to be achieved.

The balance with the proper authority of the head is clear, overlapping heavily on the boundary, with the head responsible for the operational run-ning of the school on the principles agreed. The head must educate the gov-erning body to run itself.

What governing bodies may find difficult to accept is the government's set of priorities on public expenditure. Governors generally feel that schools are under-funded, and they noted the sudden (relative) flood of additional money in the GM sector. But even if there is to be relatively less money overall, then the governing bodies will want:

- the money distributed equitably
- the government to listen
- the government to be honest about difficulties and its (relative) failures
- the governing bodies to make the local decisions.

Some governors will be tempted to quit if the government can't or won't give what those governors believe is really needed. The majority, however, will see that local management is even more important in times of pain. *'We'll* decide who has to be made redundant. We know. We can bring parents and staff along with us, given time. *If* quality *has* to suffer, we're best placed to judge for these children how quality can suffer least', commented one governor recently.

'Our job is to make sure that things are done decently, and in order', said Junior Schools Minister, Eric Forth, on 'Woman's Hour' – this is a perfectly proper definition of the role of the state in education. 'If only that's all you would do', we say on behalf of heads and governors. 'If you'd see us, heads and governors, as *allies*, it would be so much better for the government's own interests. You *need* us, to get it done in good order. Just find the government's boundary for governing the education system, and stay there. From there, negotiate with us through a new consultative policy-making forum. Just govern, stop managing.'

As for 25,000 governing bodies, we're urging you to believe in your own creativity on behalf of the children in your care. You have collective authority to govern the school. Stop asking others to draw lines for you; stop asking others to make definitions and decide priorities for you. Don't give authority away. Keep your eye on the reality of what the school means for you. Negotiate, and you can't go far wrong.

# Index

Please note that there is no heading for 'governing body' (GB), and only limited entries for 'head' and 'individual governor'.

These three key concepts are implicit in almost every other heading.